Oliver Joseph Lodge

ON THE CONTROVERSY CONCERNING VOLTA'S CONTACT FORCE

Elibron Classics
www.elibron.com

Elibron Classics series.

ISBN 1-4212-6807-8 (paperback)
ISBN 1-4212-6806-X (hardcover)

This Elibron Classics Replica Edition is an unabridged facsimile
of the edition published in 1900,
London.

Reprinted from the PROCEEDINGS OF THE PHYSICAL SOCIETY OF LONDON, Vol. xvii. pt. iv. *December* 1900.

On the Controversy concerning Volta's Contact Force.

By Professor OLIVER LODGE, *D.Sc., LL.D., F.R.S.**

CHAPTER I.

Thermodynamic Arguments.

THOSE who take what I may call a metallic view of the Volta contact-force are accustomed to deny that the Peltier evolution of heat measures the local E.M.F. existing at a junction ; they assert instead that it measures the rate at which that same E.M.F. varies with temperature :—

$$\Pi = T\frac{dE}{dT}.$$

Let us examine this assertion.

An equation of this general shape is undoubtedly true, but it has not the meaning which they assign to it.

The Π may refer to a particular junction, and in certain cases does, but the E does not. The E means the total E.M.F. of the complete circuit, always.

I venture to think that this will not be disputed, if the matter be for a few moments carefully attended to. To assist attention we may go carefully over some elementary ground—so elementary that it is difficult to secure for it that amount of attention which is necessary for clearness.

The thermoelectric power of a metal A at various temperatures means the function which is plotted in an ordinary thermoelectric diagram, with temperatures as abscissæ, relatively to some standard metal (lead) taken as the line of reference or constant zero of thermoelectric power.

To determine this function a circuit has to be constructed with the metal A and the standard metal O, their junctions brought to different and known temperatures t and t', and the resulting E.M.F. (the whole E.M.F. of the circuit) observed, being measured either by the strength of current developed, or better, by a potentiometer or compensation

* The Presidential Address delivered at the Annual General Meeting, February 9, 1900.

method. Then the ratio $E/(t'-t)$ is called the thermoelectric power of the metal A at the mean temperature $\frac{1}{2}(t+t')$, and is plotted accordingly, for a considerable range; the differences $t'-t$ being naturally kept small, in case the resulting function is not linear.

This is done likewise for other metals, B, &c.; and the relative thermoelectric power of two metals at the same temperature (being $P_{AB}=P_B-P_A$) is represented by the vertical distance between the A and B lines of the diagram.

Another function of temperature is likewise represented in the diagram, not by a length but by a rectangular area, bounded on the right by the thermoelectric height, and on the left by the line of absolute zero. This is the function Π, such that

$$\Pi_{OA}=P_AT, \quad \text{and} \quad \Pi_{AB}=P_{AB}T.$$

It is to be observed that in this diagram (apart from the experimental mode of obtaining real values for P) no reference is so far made to any junctions, nor to any circuit, nor current, nor E.M.F., nor evolution of heat; it represents nothing but a geometrical plotting of certain metallic properties as functions of absolute temperature.

If we now connect the two metals at one point, they are necessarily at the same temperature, say T_1, at that point; their relative thermoelectric power at the junction, P_{ABT}, may be called P_1; and the area just spoken of is $\Pi_1=P_1T_1$.

Fig. 1.

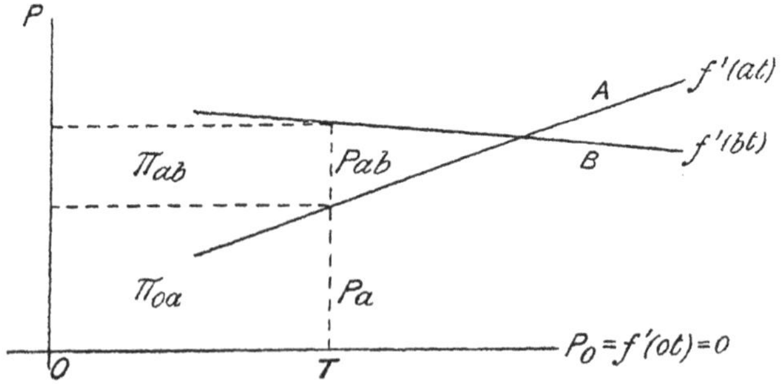

Now Lord Kelvin showed in 1856 that this Π_1 is the Peltier coefficient, *i.e.* that it measures the reversible heat generated at the junction if a unit quantity of electricity is transmitted across it.

If we proceed to connect the two metals also at some other point, at some other temperature T_2, there will of course be an appropriate value for P_2 and Π_2, but there is more than that: every intermediate point of one metal will have a temperature corresponding to that at some intermediate point of the other metal, and accordingly there is a P and a Π appropriate to every intermediate temperature.

There will also be now a closed circuit, and a current, driven by a resultant electromotive force, which may be expressed instructively in the following different ways,

$$E_{AB} = \text{cycle}\int P dt = \int_1^2 P_{AB} dt = \int_1^2 \Pi d \log T = f(t_2) - f(t_1) \;;$$

the function f being such that

$$\Pi_{AB} = T f'(t) = T \frac{d}{dt} E_{AB} \text{ everywhere.}$$

At the junctions there is now a reversible generation of heat per second

$$\Pi_1 = T_1 f'(t_1) \quad \text{and} \quad \Pi_2 = T_2 f'(t_2)$$

per unit current.

In the metals there is likewise a reversible generation of heat

$$\Theta_a = \int_1^2 \sigma_a dt, \qquad \Theta_b = \int_1^2 \sigma_b dt,$$

such that

$$\sigma_a - \sigma_b = T \frac{d^2 E}{dT^2} = T f''(t) \;;$$

and the total E.M.F., in addition to the modes of writing adopted above, may also be represented as,

$$E = J(\Pi_1 - \Pi_2 + \Theta_a - \Theta_b),$$

since E represents the work done per unit quantity of electricity conveyed.

The representation of all this in a thermoelectric diagram is neat and convenient, and is sufficiently shown in the figure, without overlapping of areas.

Remembering that it is P which is plotted vertically, and that $P=\Pi/T$ everywhere, the whole of the above various expressions for the area E are geometrically represented.

Fig. 2.

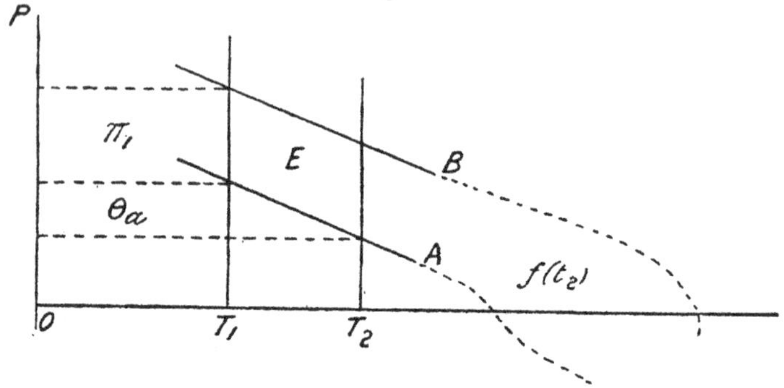

The only one needing explanation is the expression $f(t_1)-f(t_2)$, the difference of two indefinite-integral areas, of which the second limit is incomplete in the figure, because unknown. All that is known about $f(t)$ is that E is the difference between its values at the temperatures t_1 and t_2, that $f'(t)=\Pi_{AB}/T$ everywhere, and that $f''(t)=(\sigma_a-\sigma_b)/T$ everywhere. But the value of the function $f(t)$ itself is not known; nor is there any means of determining it from observed values of a current in a closed thermoelectric circuit. (I have no reason for depicting this area in the diagram on the right towards infinity, instead of on the left towards absolute zero, except that thereby overlap is avoided. Otherwise one would naturally draw it so as to be bounded hypothetically by the line of absolute zero. The diagram does not attempt to represent its probable value, but merely to emphasize the fact that its absolute value is not known or needed in ordinary thermoelectricity.)

Excepting in portions relating to $f(t)$ there is, I apprehend, so far no controversy. And yet, if all this be admitted, my contention that, in the equation $\Pi=TdE/dt$, the E refers to the whole circuit and not to any particular part of it, though the Π *may* refer to a junction, is substantiated.

Nevertheless this statement has been, and probably still

will be, seriously disputed *, but I hope that if it is still disputed the actual subject will be dealt with, and not hydraulic illustrations of doubtful applicability given.

For instance, I might quote several physicists who have said (somewhat casually, as I think), since a discussion at the Institute of Electrica Engineers (Journ. Inst. E. E. 1885), that the Peltier heat-evolution is no measure of a contact E.M.F. existing at the junction where it occurs, but is a measure of the rate of variation of this E.M.F. with temperature; but it will suffice if I quote part of Prof. Ayrton's remark in that same discussion, with an occasional interposition of my own in square brackets, because he expresses himself clearly and strongly—always a desirable thing to do:—

"Dr. Lodge's fundamental argument—the argument in fact upon which his whole paper is based—is this: He *assumes* that in all cases where there is a considerable difference of potential at a [metallic] junction, heat or cold must be developed when a current flows across that junction." [I should myself word it rather differently, to avoid misunderstanding, though misunderstanding is hardly possible in what follows; but apart from possible misunderstanding I have no objection to the statement, except that I do not regard it as an *assumption*, unless the conservation of energy is an assumption.] "He states what is quite true and well known—that practically no heat is developed when a current flows from zinc to copper across the junction; therefore he concludes that there can be no contact-potential difference at this junction like three-quarters of a volt. But this reasoning is based on a totally wrong conception of the Peltier effect, which I am astonished to find exists in the mind of a man with the scientific powers of Dr. Lodge. *The amount of heat generated when a unit current is sent through a junction for a second is undoubtedly the measure of the coefficient of the Peltier effect; but this coefficient is in no sense a measure of the contact-potential difference existing at the junction.*" [Italics are Prof. Ayrton's.] "What the Peltier effect, at a junction of the two substances at a given temperature, measures, is the product of the absolute temperature into the rate of variation, with temperature, of the contact-potential difference at that temperature; in other words, if Π is the Peltier effect at a junction which is at an absolute temperature t, and if V is the contact-potential difference at that temperature, $\Pi = t\frac{dV}{dt}$.

"Hence the magnitude of Π does not measure V, but merely $t\frac{dV}{dt}$, and the error of Dr. Lodge is of the same sort of order as saying that a train going at a very nearly uniform speed is necessarily moving very slowly because the change of speed per second is very small. This disposes of the objection based on the smallness of the Peltier effect." [No, it is not disposed of quite so easily as that. The difference between velocity and acceleration is somewhat familiar.]

Controversy begins as soon as localization is attempted; as soon as one says that $J\Pi_1$ is a measure of the E.M.F. existing at the junction whose temperature is T_1, and that $J\Theta_a$ is a measure of the E.M.F. existing along the gradient of temperature in the metal A.

Controversy also begins in respect of the function $f(t)$ *. There are those who decline to take an agnostic position with respect to the absolute value, or second limit, of this function, and who say that it represents absolutely the Volta contact force; that is, the E.M.F. observed in the dielectric space between two metals when a metallic circuit is not completed, when there is only one metallic junction, and when a dielectric space is substituted for the other junction. The value of the E.M.F. thus observed is very large, say a thousand times as big as any we have been so far dealing with; it is not observable in any closed metallic circuit, it is only observable in a dielectric portion of a circuit; and it has no apparent connexion with thermoelectricity; hence, those who assert that this great E.M.F. is represented by $f(t)$, that is, by a function of the metallic junctions otherwise and solely known by its differences and derivatives in thermoelectricity, should be prepared with a definite proof of their statement. So far as I know a proof has never even been attempted. To me it appears that the introduction of a dielectric or an electrolyte into the circuit, with its necessary two new junctions, each with chemical potentiality, has entirely altered the whole conditions; and that no longer can it be said that the above laws, appropriate to a closed metallic circuit, apply. I maintain that whether the assumption, that $f(t)$, a function of the metallic junctions alone, represents the Volta effect observed in the air when a circuit is broken, is made tacitly or explicitly, it is made gratuitously and requires justification, which, so far as I know, has not yet been forthcoming.

To show that the unsupported statement just referred to —whereby the Volta effect, a potential gradient observed in the air when a metallic circuit is broken, is introduced into a thermoelectric function of metals alone, with which it appears to me to have nothing to do—is responsibly made, I might quote from several writers; but there is no

* *Cf.* Phil. Mag. March 1886, p. 267.

need to quote more than Lord Kelvin's statements in his Friday evening discourse to the Royal Institution for May 1897 (Reprinted in the Phil. Mag. July 1898) :—

"§ 19. Seebeck's great discovery of thermoelectricity (1821) was a very important illustration and extension of the twenty years' earlier discovery of the contact-electricity of dry metals by Volta. It proved independently of all disturbing conditions that the difference of potentials between two metals in contact varies with the temperature of the junction." [This is no doubt true as it stands, if the junction is the only part heated; the Seebeck force is superposed upon the Volta force, and hence the electrostatic effect observable in an incomplete circuit is naturally slightly affected by differences of temperature, though not necessarily by absolute uniform temperature. The break of circuit may be made in the middle of one of the metals whose junctions are maintained at different temperatures, in which case only the Seebeck force could possibly be electrostatically observed (that is, a combination of Peltier and Thomson effects) by a sufficiently sensitive electrometer; or the break may be made between the two metals, in which case the much greater Volta force will be observed in the air-gap in combination with the insignificant Peltier force of the metallic junction; and any fluctuation of temperature to which that remaining metallic junction is subjected will but vary this insignificant superposition upon the true Volta effect. But the sentence quoted is not intended to convey this meaning; what it is intended to convey is that the Volta gradient of potential observed in the gap is chiefly due to something occurring at the remaining metallic junction, and that this something is a function of the temperature. Indeed, a later sentence (§ 26) clearly states, not only the existence of this large metallic-junction force, but also its cause.] "This force" [viz. the force of attraction between the oppositely-charged metals across a gap], "properly viewed, is a resultant of chemical affinity between thin surface-layers of the two metals."

[And then the author goes on to explain that the junction-force at the boundary of two metals has nothing whatever to do with any reversible heat-effects which may be observed there; because, if it had, these reversible heat-effects would be vastly larger than in fact they are. It is therefore argued that these Peltier effects by no means represent the value of the E.M.F. existing at the junction where they occur, but represent only the rate at which this E.M.F. varies with temperature, $\Pi = \mathrm{T}d\mathrm{E}/d\mathrm{T}$; whereas, as I have said above, the E really appropriate to this equation is not located at the junction at all, but is the resultant or integrated E.M.F. of the whole circuit. However, I continue the quotations, not at first using quotation marks simply because the wording is abbreviated where no precision is required.]

Imagine a circuit of two metals—say iron and copper; and let the copper be cut across in the middle, so that an independent E.M.F. can be introduced by wire electrodes and a current maintained across the two junctions. "The Peltier heat produced where the current passes from iron to copper is manifestly not the thermal equivalent of the work-done." [This sentence is ambiguous. It may mean the "work done" on the *whole circuit*, in which case it is true; it may mean the "work done" precisely *at the junction*, in which case I should say that the word "not" ought to be omitted. The immediate context suggests the former interpretation; the next succeeding context suggests the latter. The immediate context runs as follows:]

"In fact, if the two junctions be at equal temperatures, the amounts of Peltier heat produced and absorbed at the two junctions will be equal, and the work done by the independent E.M.F. will be spent solely in the frictional generation of heat." [This is undeniable, though not apparently to the point; but the next section continues]

"Many recent writers, overlooking . . . obvious principles, have assumed that the Peltier evolution of heat is the thermal equivalent of E.M.F. at the junction. And in consequence much confusion, in respect to Volta's contact-electricity and its relation to thermoelectric currents, has clouded the views of teachers and students. We find over and over again the statement that thermoelectric E.M.F. is very much smaller than the Volta contact E.M.F. of dry metals. The truth is, Volta E.M.F. is found between metals all of one temperature, and is reckoned in volts, or fractions of a volt, without reference to temperature." [Certainly.] "If it varies with temperature, its *variations* may be stated in fractions of a volt per degree." [By all means.] "On the other hand, thermoelectric E.M.F. depends essentially on *difference* of temperature" [the thermoelectric force of a complete metallic circuit does, and not the thermoelectric force at any particular junction; that depends on the temperature of the junction, and cannot be affected by the temperature somewhere else] "and is essentially to be reckoned *per degree*; as, for example, in fraction of a volt per degree." [Not so; an electromotive force of any kind whatever must be reckoned in volts. Volts per degree is not E.M.F., but is the so-called relative thermoelectric power of two metals.]

I have ventured thus to interpolate what I consider corrective remarks in brackets into this readily accessible quotation, because of a preceding sentence with which I quite agree, viz. this:—"Peltier's admirable discovery . . . is highly important in theory, or attempts at theory, of the contact electricity of metals"*.

* To this I might add, what I now see more clearly than before, that it is more particularly the mode of statement of Lord Kelvin's own "admirable discovery" that is really the central feature of this part of the controversy. For there are three, at least three, modes of regarding it:—

1st. A convection of heat pure and simple—electricity acting as a fluid with a positive or negative specific heat, and conveying heat down or up a temperature gradient.

2nd. An E.M.F. acting in metal from hot to cold or from cold to hot, and thereby either opposing or assisting the passage of a current, in the one case developing extra local heat, in the other case consuming heat or developing local cold, proportional to the local E.M.F. and the current.

And this 2nd mode has two subdivisions:—The E.M.F. thus supposed to exist from end to end of a metal bar with its ends at different temperatures, may display itself:—

either (*a*) as a gradient of potential in the air outside, detectable by an abnormally sensitive electrometer-needle, but never yet observed;

or (*b*) as a change in the step of potential from metal to air, according to the temperature of the metal; the air-potential outside being left uniform.

But there is yet the third method of regarding the whole matter:—

And in case anything that I can say can make the matter clearer, I would reassert that since E.M.F. is work per unit charge conveyed, not only for the whole circuit but for any

---

3rd. As a mere modification of the junction E.M.F. with temperature; so that the Peltier function, which had been of the value $kt(t_0-t)$, is modified by addition of a term $\frac{1}{2}kt^2$, so that it becomes $kt(t_0-\frac{1}{2}t)$.

On this plan, no E.M.F. of any kind is supposed to exist in the single metal itself, no more than when its temperature was uniform; all the E.M.F. is thrown upon the junctions; the E.M.F. of the circuit is simply the difference of the junction-forces, as it would have been had there been no Thomson effect; but each junction-force, by reason of the Thomson effect, has somehow attained a quadratic term; their difference therefore permits thermoelectric reversals and the observed law of the thermoelectric circuit, for subtracting the two junction-forces we get

$$E=k(t_1-t_2)(t_0-\bar{t}).$$

This third mode is evidently the way in which several philosophers have been accustomed to regard the matter; and on the hypothesis that junction-forces in a metallic circuit are due to the affinity of metals for each other, it is natural to locate all the E.M.F. at the junction of two different metals, and to deny any E.M.F. between different parts of one and the same metal at different temperatures. (This is very different from the experimental fact that the resultant E.M.F. round a homogeneous metallic circuit is zero.) Those who regard the junction-forces as physical would naturally take a different view. For instance, take Helmholtz's hypothesis of a specific attraction of metals for electricity, the same being a function of temperature: a differential attraction between the hot and cold ends of a bar would be just as essential as a differential attraction at the junction of metals of different kinds. It has been asserted that Helmholtz's view tends to concentrate the force at the junctions and to make the Peltier heat proportional to the rate of variation of the local junction-force with temperature. But even if it does, and later we will show that it does not, no view makes this junction-force the same as the Volta effect, or suggests that the Volta effect has anything to do with the matter; except on the hypothesis that the air layer intervening when a metallic junction is broken is utterly inert and impotent.

It is no argument in favour of the impotence of the air layer to say that it produces no apparent effect when it bathes the metal uniformly all round; the question is what it does when uniformity all round does not exist. Balanced forces are not the same thing as no force, although they produce the same zero accelerative effect. Nor does the absence of acceleration prove the absence of all force. A body subject to balanced forces and obeying the first law of motion is a frequent spectacle—a railway train or a snowflake for instance—but a body subject to no forces is decidedly rare.

portion of a circuit, and since a unit charge conveyed across a junction develops an amount of heat $\pm\Pi$, it follows that the physical E.M.F. located at that junction is measured by and is equal to $\mp J\Pi$ (assisting the current in one direction and thereby consuming heat or developing cold, opposing it in the other direction and therefore developing extra heat); unless some other mode of accounting for the appearance or non-appearance of the reversible heat is devised, or unless the reversible energy takes some other form than heat.

Wherever energy appears or disappears in a circuit, there, in that place, must be located an E.M.F. It may be, so to speak, the frictional E.M.F. appropriate to Ohm's law in a homogeneous conductor; it may be the chemical E.M.F. at a fluid or semifluid junction discovered by Volta; or it may be the contact-force at a solid junction, of which the integral round a whole circuit of non-uniform temperature was discovered by Seebeck, the locality of part of that E.M.F. by Peltier, and the locality and existence of the rest of that E.M.F. by Lord Kelvin.

At a chemical junction the reversible energy need not all of it or much of it take the form of heat, and in that case the major part of the E.M.F. may be rightly called chemical, not physical; but at a metallic junction, as in a homogeneous conductor, heat is usually the only form of energy expected or observed, and in that case wherever the E.M.F. is located there the heat or cold must appear. The ordinary way out of this conclusion is to suppose that electricity is a fluid with a real specific heat whose value changes from metal to metal, so that heat can be evolved or absorbed at a junction by a sudden change in the heat-capacity of the electric fluid flowing across it, without the need for any local performance of work or any corresponding local E.M.F.*

I should be glad to know whether any objectors to what I may call the Maxwellian view—the view of the precise local correspondence between E.M.F. and reversible heat-effects—do seriously base their opposition on this ground of a real physical specific heat of electricity. The only other alternative that I can see (short of Helmholtz's theory, to be mentioned later) is to say that the metallic junction is

* *Cf.* Phil. Mag. March 1886, pp. 271-276.

really a chemical junction too, that the energy does not take the form of heat because it takes that of chemical combination, and to point to a weighable quantity of alloy or other compound produced at that place by the passage of a current. Whether or not any chemical compounds are really formed at a given junction, then becomes a simple question of fact.

Since a voltaic circuit never has less than three junctions, and since we are trying to apply to it as much of the thermoelectric laws as it will stand, it may be well to write down these laws for the case of a metallic circuit with any number of metals, and therefore with a thermoelectric diagram having discontinuities at the several junctions,—everything discontinuous but temperature. Each metal, referred to the standard, has a certain thermoelectric power at each temperature, and accordingly the resultant E.M.F. of the circuit is

$$E = \text{cycle}\int P\,dt.$$

We may also express the E as a sum of the E.M.F.'s at the several localities, the junctions on the one hand, and the temperature-gradients in the metals on the other; so that

$$E = \Sigma(\Pi) + \Sigma(\Theta).$$

Further, as regards the Π function of each metal relatively to the standard

$$\Pi_{oa} = P_a T \text{ everywhere};$$

and also

$$\Theta_a = \int_1^2 \sigma_a dt;$$

so, maintaining this reference to the standard, it is still true that

$$E = \Sigma\int \Pi\, d\log T.$$

But Π_{ab}, though still equal to $P_{ab}T$, is no longer equal to TdE/dT, unless by E is meant E_{ab}, that is the E.M.F. of a non-existent small portion of circuit formed solely of the two metals A and B.

The fact is, that at a junction there are now three values of P and of Π, viz. Π_{oa}, Π_{ob}, Π_{ab}; and it is only when we constantly maintain the reference to the standard metal o that we can say $E = \Sigma\int \Pi\, d\log T$; otherwise there are discontinuities not allowed for. We cannot say that

$$\int_1^2 \frac{\Pi_{za}}{T}dt + \int_2^3 \frac{\Pi_{ab}}{T}dt + \int_3^4 \frac{\Pi_{bc}}{T}dt + \ldots + \int_n^1 \frac{\Pi_{yz}}{T}dt$$

is equal to E. For one thing, the temperature-limits and the junctions do not fit. An expression of the form $\int_{t_1}^{t_m} \frac{\Pi}{T}dt$ can still be employed for E when care is taken to arrange the Π/T so as to represent the difference in thermoelectric powers between pairs of points of the circuit which are *at the same temperature*, whether the metals possessing that temperature happen to be anywhere directly in contact or not. With this precaution, the integration can proceed over a discontinuity at every junction; but, in order to select the appropriate temperature-limits, either reference must be made to the real temperature of different parts of the circuit, or else the simple plan of referring everything to a hypothetical standard metal lying along the axis of temperature must be employed.

This being understood, we can still write $\Pi/\mathrm{T}=d\mathrm{E}/d\mathrm{T}$; but, while the E refers to the whole circuit, the Π does not now refer, even permissibly, to the junctions; except the hottest and coldest junctions; unless reference is made continually to the standard metal (some point of which is necessarily at the same temperature as any given junction), and not to the particular pair of metals which happen there to touch.

Hence we may not assert of the actual junctions at different temperatures that $\Sigma(\Pi)=\Sigma(\mathrm{T}d\mathrm{E}/dt)$; nor can we assert without caution that $\Pi=\mathrm{T}d\mathrm{E}/dt$ at each junction individually.

So far for a multiple metallic circuit.

If the circuit is not wholly metallic, but is partly dielectric or electrolytic, the introduction of the extraneous substance complicates matters. There may be a resultant E.M.F. now without any difference of temperature; and this E.M.F. may be partly chemical and partly physical, partly $\Sigma(\mathrm{J}\theta\epsilon)$, to use the customary notation (Kelvin, 1851), and partly $\Sigma(\mathrm{J}\Pi)$. For even if the temperature is everywhere uniform, it is no longer necessary for the Π's to balance each other; all one can be sure of is that when there is no gradient of temperature there can be no Thomson force Θ; and that the resultant E.M.F. is therefore the algebraic sum of the junction forces,

$$\mathrm{E} = \Sigma(\mathrm{J}\Pi) + \Sigma(\mathrm{J}\theta\epsilon).$$

It cannot now be assumed without further special proof that at any given junction $\Pi=\mathrm{T}d\mathrm{E}/d\mathrm{T}$; nor can we assume that it is permissible to individualise the junctions; but still, for an electrolytically closed circuit at uniform temperature, von Helmholtz has shown that taking *all* the junctions into account,

$$\Sigma(\Pi)=\mathrm{T}\frac{d\mathrm{E}}{d\mathrm{T}},$$

the E being the resultant E.M.F., the sum of the contact-forces at *all* the junctions whether they be of a chemical or physical nature. And recently (Proc. Roy. Soc. Edin. Feb. 7, 1898, vol. xxii. p. 118, "On the Thermodynamics of Volta

Contact Electricity") Lord Kelvin has proved exactly the same thing to hold for a dielectrically closed circuit of uniform temperature likewise. The argument in this case he considers quite rigorous, because there is now no questionable gradient of temperature along heat-conductors, and therefore no necessarily concomitant irreversible effect; and therefore no question of whether such effect be accidental or essential.

The only loophole I can perceive for a flaw in the reasoning is the possibility that the heat of chemical combination varies for small temperature changes, *i. e.* that it is a perceptible function of the temperature; and I have no reason for pressing this, or for hesitating over accepting the above conclusion to the utmost, so long as the temperature is uniform.

It is unimportant for the present argument, but it may be worth while to notice in passing, that no such relation as $\Sigma(\Pi)=\Sigma(\mathrm{T}d\mathrm{E}/dt)$ has ever been proved for a complex circuit in which differences of temperature are superposed upon other sources of electromotive force. And from what has been said in similar small type just above, no such relation can truly be asserted, without considerable care as to the form of the statement. Its most obvious signification, having reference to the actual junctions, would not be correct. It is to be observed that in the Kelvin and Helmholtz expression for a complex circuit at uniform temperature, the meaning of $d\mathrm{E}/dt$ is a change of E.M.F. of the circuit per degree caused by the passage of a current—caused, that is to say, by a thermal polarization: the dt/T being the efficiency of a cyclical Carnot engine. Whereas in such an expression in ordinary thermoelectricity as $d\mathrm{E}/dt=\mathrm{P}=\Pi/\mathrm{T}$, the meaning of $d\mathrm{E}$ is the total E.M.F. observed in a circuit of two metals when its pair of junctions are artificially maintained at a difference of temperature dt.

It may be said that these two meanings are after all the same: and so they are in a simple circuit of two metals, but not in a complex circuit with junctions at more than two temperatures.

But what has all this to do with the localisation of E.M.F. at a given junction? Admitting, and indeed teaching, all this, I still hold that the reversible heat at a specified junction is a measure of the metallic E.M.F. located there. That in fact at a metallic junction $e=\Pi$, or $\mathrm{J}\Pi$ if anyone thinks it worth while in a matter of this sort to trouble about practical laboratory units.

Those to whom I referred in the opening paragraph of this communication make the mistake of confusing e with E,—confusing the E.M.F. at any one junction with the whole E.M.F. in a circuit. If they do this, they naturally beg the question, and of course locate the whole E.M.F. at the particular junction which takes their fancy at the moment: usually an interface of zinc and copper.

The truth is that $E = \Sigma(e)$ always; but only in a metallic circuit at constant temperature does $\Sigma(e) = \Sigma(\Pi)$; just as only at a metallic junction does $e = \Pi$ individually. At an electrolytic or dielectric or what I may call a chemical junction, that is one with chemical potentialities, the E.M.F. is not purely thermal, and hence is not measured by the Peltier effect; it is chiefly of chemical origin and is calculable from the energy of combination of the materials on either side the boundary.

Observe I do not say that at such a junction there need be any chemical *action* (at least not any completed chemical combination in the ordinary sense) to account for the E.M.F., but there is the possibility of chemical combination; and there will be actual chemical combination if a finite quantity of electricity is transmitted across it. For conduction in an electrolyte is convective in its nature; whatever electricity travels, the material atoms travel with it; and the travelling of oppositely charged ions in opposite directions involves chemical action at a liquid junction, involves transfusion of substance, involves what is practically, and may be actually, chemical recombination. Possibly something of the same sort may be said, less positively, about electric displacement at a dielectric boundary.

But at a metallic junction there is no such chemical potentiality. You may pass a strong current across a zinc-copper junction for years and you will not get any brass. I would not deny that there may be a kind of material or "corpuscular" interchange even here: in fact in the light of recent most interesting work by J. J. Thomson it is probable; but the corpuscles thrown across the junction from zinc become copper; and *vice versa*, those dissociated from a copper atom settle into their new places as part of a zinc atom; there may thus be a tendency to an ineffective kind of *transmutation*, but there is no tendency to alloy. (See below, p. 420.)

Why then if a metallic junction is chemically inert should it ever be held that the chemical affinity of zinc for copper is the propelling influence which causes the E.M.F. located at such a junction? What has ordinary chemistry got to do with this metallic E.M.F.? I echo the question, and proceed to my second chapter.

Chapter II.

The facts of Contact Electricity.

Some people (for whom I have a great respect), on the strength of the fact that metals in contact attract each other, maintain that this accounts for their tendency to combine and for their heats of combination, as well as for the value of their contact E.M.F.

Now, that metals in contact attract each other I fully admit. They must, because they are oppositely charged. There was a time when people disputed the *fact* of the Volta effect, but I am not one of those, and never was. All the facts by which Lord Kelvin or anyone else properly displays the Volta effect I fully admit, and am accustomed to show to my class; though indeed in Lord Kelvin's Friday evening Royal Institution discourse (Phil. Mag. July 1898) they are described anew with admirable clearness and precision. To avoid any possible mistake on this score, I will briefly recapitulate the facts.

Dry zinc and copper in dry air brought into contact become oppositely charged. If their surfaces are arranged so as to form a condenser of reasonable capacity, the charges may be large. When they are separated adroitly, the charges remain on the metals and raise them to a high (numerical) potential which may readily be displayed by an electrometer or even a gold-leaf electroscope.

A trace of moisture, *i. e.* of liquid electrolytic moisture, not dry vapour, may be fatal to the success of this experiment; for if a drop of liquid intervenes and connects the two metals momentarily, after the true metallic contact has been broken, nearly all their charge leaks across this conducting-bridge; and the resulting charge, and therefore the resulting potential of the separated plates, is extremely small. A water junction between two plates has no appreciable power of charging them oppositely; practically no more power than air has. In many respects a water intervention acts very like an air intervention; in one respect only does it strikingly differ:—its conductivity enables it to abolish any effective display of the opposite charges otherwise produced, *e.g.* by metallic contact. The same detrimental influence is exerted by the conductivity induced in air subjected to the action of *x*- or of uranium rays; such air behaves precisely like a badly conducting electrolyte. Nothing further need be said to make clear the action of metallic uranium for present purposes.

It is remarkable, in practice, how imperceptibly minute the quantity of liquid may be, in order thus to act detrimentally. But after all it is only a question of arithmetic:—

Let z be the average thickness of dielectric intervening between the zinc-copper plates, during their so-called close contact, everywhere except at the place where they really effectively, *i.e.* metallically, last touch as they are being separated; and let z' be the distance between the same surfaces at the instant when the hypothetical moist or liquid connexion is likewise last broken; then in the dry case the resultant charge is inversely proportional to z, being in fact of the order $\frac{r^2}{1200z}$ electrostatic units, for the usual case of circular disks; whereas in the moist case z' must be substituted for z.

Now a very minute drop of liquid pulls out before it snaps into a column π tenths of a millimetre long, which then represents z'; whereas the dry z may be a tenth or even a hundredth of this magnitude, with well-made plates and careful manipulation.

I have emphasised this to negative any possible idea that I regard moisture as an adverse fetish, acting discontinuously. It is better away; but, if it exists, its detrimental effect is a mere arithmetical affair.

No one, I hope, can now suppose that I imagine that moisture is essential to the success of the experiment, though in his Friday evening discourse to the Royal Institution, May 1897, reprinted in the Phil. Mag. (July 1898), Lord

Kelvin says that there are signs in my British Association Report (Phil. Mag. 1885) of a tendency to fall back upon De la Rive's old [and utterly discredited] hypothesis that the Volta effect was due to oxidation of the zinc by moisture from the air. He means, I expect, that I attach too much importance to the action of oxygen (which may be true: a word on that later), but never have I imagined it to be needed in the form of moisture, nor have I ever wanted the zinc to be actually oxidised. If it be perceptibly oxidised, the Volta effect is diminished.

I say once more—in accordance, I believe, with everybody:—

(1) Two metals connected wholly or partly by liquid are in approximately the same electrical condition as if connected wholly or partly by air. They do not become oppositely charged unless they are at one (or more) point metallically connected: then they do: and in order to be able fully to display this charge on separation, the persistently-connecting medium must be insulating, that is must be dry, *i.e.* free from liquid. It must also be free from high-frequency ætherial radiation or other ionising influence.

(2) If the metals are connected *both* metallically and electrolytically, metallically at one place and electrolytically at another, we have a common voltaic cell and a continuous current.

(3) To complete the statement:—If the metals are connected both metallically and *dielectrically*, that is if they are insulated from each other except at one point where the intervening dielectric is swept away, we have the normal conditions appropriate for a display of the Volta effect: that is, a gradient of potential in the dielectric between them.

The safest and clearest mode of expression is what I have used: the Volta effect consists in an opposite charge acquired by dry zinc and copper while in metallic contact, a charge which results from an E.M.F. of value depending on the condition of their outer surfaces, and controlled solely by this E.M.F. and the electrostatic capacity.

I here avoid any statement involving the term metallic potential, because it is open to misconception owing to variety of definition. Some people prefer to say that zinc and copper in contact are at different potentials, whereas I prefer to say that they are at the same potential so long as they are in contact: but no one can deny that they are oppositely charged.

Some people, again, prefer to say that they are oppositely charged by reason of an E.M.F. at the junction, which keeps them at different potentials: I prefer to say that their charges are distributed so as to enable them to remain at the same potential, in spite of a difference in the specific Volta force existing at the dielectric boundary of each; in other words, that the E.M.F. is at their surface; but again, no one can deny that there exists in the dielectric between them a gradient of potential, the natural result of their opposite charges; which same gradient can cause a charged electrometer needle to move, even without the magnification of potential caused by separation. In fact the magnification of potential by separation does not (within limits) increase the *gradient*, for the potential rises as the capacity diminishes, and so remains in direct proportion to their distance apart, so long as that distance is moderate.

Nor can anyone deny that oppositely charged bodies, or surfaces between which there is a gradient of potential, attract each other. Hence, indirectly, a piece of zinc attracts a piece of copper with which it is at any point in contact; and the attraction increases as the square of the potential gradient, and therefore as the square of the nearness together of the opposed or condenser surfaces.

Signor Majorana (Phil. Mag. Sept. 1899) has designed some simple methods for displaying this attraction, one of them the same as a method I had used for displaying the attraction of two metals exposed to Hertz radiation just before cohesion; viz., a fine metallic fibre hanging in front of a polished facet on a knob, the fibre and its image being watched through a microscope (Discourse on Coherers, Journ. Roy. Inst. 1899): but what has all this to do with the location of contact force at one junction rather than at another? What bearing has it upon the present controversy? So far as I see, none.

And further, I would ask, what has this dielectric attraction got to do with the heat of formation of brass? But in asking this last question I do not pretend that it is an easy one, and I would not dogmatise as to the answer. The question is briefly treated at the end of the present chapter. I only wish to indicate at this stage that the answer to the question is open to doubt, and that it must not be taken as indisputable that the heat of combination of metals has any connexion with their voltaic force, or with their thermoelectric force either.

There was a time when I fancied that though the heat of formation of brass had no connexion with the Volta force, yet it might have some connexion with the thermoelectric force of contact, or, let us say, the Peltier effect. This idea, too, I wish to abandon*, and for the same reason, viz., that if metallic affinity were an efficient agent in propelling the current, then I should expect the transmission of the current to result in some formation of alloy. Conceivably some secondary influence may mask the display of this result; and if the numerical determinations of metallic heats of combination correspond, when reduced to volts, with the E.M.F. anyhow observed in connexion with circuits or partial circuits formed of those metals, that would be a striking coincidence and one to be carefully considered. So far as I am aware, no such coincidence has yet been *indisputably* discovered, even with a single pair of metals.

Experiments on heats of metallic combination have been made by Sir W. Roberts-Austen, by Dr. Galt, by Mr. T. J. Baker, and by Mr. J. B. Tayler. Doubtless also by other workers whom I forget at the moment or do not know of. The results obtained are not yet free from doubt and difference of opinion. Some of the results agree neither with the Volta nor the thermoelectric forces; and this discordance is what I expect will ultimately be believed. On the other hand, some determinations of the heat of alloying of zinc and copper, taken in conjunction with the customary estimate of the size of atoms, do appear approximately to coincide with the order of magnitude of the Volta contact-force. (See end of this chapter.)

If this is truly so, it is very remarkable, and must greatly strengthen the ground of those commonly called "contact theorists." It would seem unlikely to be the result solely of accident. On the other hand, it is undeniable that (either accidentally or otherwise) the order of magnitude of the Volta

* *Cf.* Phil. Mag. 1885, p. 356, and Journ. Inst. E. E. xiv. 1885, p. 219.

force can be calculated from the differential heats of combination of the metals for oxygen; which may be written (Zn, O) − (Cu, O). Hence, if it turns out that it can likewise be calculated from the heat of formation of brass (Zn, Cu), it will be either an odd coincidence or it will have an important chemical meaning with regard to heats of combination or chemical affinities in general.

One way of putting its meaning may be that it tends to strengthen what I may call von Helmholtz's hypothesis *, which can be stated roughly thus :—That the metals do not really attract oxygen, nor do they really attract each other, but that what they attract is electricity; that all chemical attraction, apparently of matter for matter, is the consequence of this real attraction of matter for electricity, the atoms being electrically charged. Further, that the Volta effect is caused, simply enough, by the differential attraction of zinc and copper for electricity. The orthodox way of putting it is as follows :—The zinc attracts positive electricity most; hence, when they are put into contact, electricity flows across the junction and the zinc becomes positive and the copper negative. Another way of putting it is to say that the zinc free surface attracts negative electricity most, and accordingly pulls it in from the space around as soon as the *uniformity* of the force is anywhere interfered with by an inert metallic contact. In either case its ready oxidisability by negative ions is thus explained, and has not to be taken as an independent fact or basis. Whereas, then, I had proposed to calculate the Volta effect as (Zn, O) − (Cu, O), on this view it would have to be reckoned as (Zn, E) − (Cu, E); though what is to be considered precisely as meant by E may be matter for further consideration. I am not going to say anything against this far-reaching hypothesis in a hurry. It demands separate treatment in the light of 'electron' theory.

This is what I intended when I admitted it possible that, as Lord Kelvin said, I had too great a veneration for the need for oxygen in this connexion. The sole reason why I think that it is probably the oxygen film and not the æther discontinuity which causes the force, is that its magnitude corresponds so closely, not only in order of magnitude but in

* Summarised in Phil. Mag. October 1885, p. 378.

relative details, with the appropriate energy of oxidation. At the same time, if the electrical attraction is the *cause* of oxidation tendency, this quantitative relation is not likely to afford any means of discrimination.

The evidence that leads Lord Kelvin to the conclusion that the æther or space or vacuum is just as effective (or ineffective) as oxygen, is derived from such experiments as those of Bottomley (see also von Zahn, &c.) to the effect that the Volta force continues unchanged in a vacuum : that an alteration of the *metals* affects its value largely, but that a change of the surrounding *medium* does not matter, unless it actually corrodes the metals and spoils them.

But this evidence derived from experiments on vacua has never seemed to me conclusive, because nothing like a real vacuum is ever attained, and even if it were, it is known with what surprising tenacity ancient air-films cling to solid surfaces ; the gaseous substance within molecular range of the solid substance being almost a part of it and not readily changed. It clings by cohesion, just as the parts of the solid substance itself cling. It is subject to Laplace's K, enormous molecular pressure. Mere exhaustion, lessening the pressure by one atmosphere, hardly disturbs it.

It was the extreme difficulty of making experiments in vacuum and substituted-media crucial that long ago deterred me from attempting them. Mr. J. Brown attempted some * ; and now fortunately Mr. Spiers has made a brave and distinctly successful attempt in the same direction, an attempt that will, I hope, be persisted in †.

That the free or bounding surface of the metal has something to do with the Volta effect, at least in modifying its value, is proved not only by the long-known observations, say, of Ayrton & Perry, of Pellat and others, on the effect of corrosion and oxide films, but also by the interesting experiments of Mr. Erskine Murray on burnishing ‡. Simple burnishing of the parts *not* in contact can run the Volta force up considerably, from ·7 volt to 1·02 volt in a certain instance;

* Proc. Roy. Soc. vol. lxiv. p. 369; see also vol. xli. p. 294, and Phil. Mag. 1878, 1879, and 1881.

† Phil. Mag. January 1900.

‡ Proc. Roy. Soc. vol. lxiii. p. 113.

and scratching drops it again*. But these surface-effects are admitted by Lord Kelvin, and are not thought to be even indirectly concerned with the zinc-copper junction and the contact-force there located†. I should say that so great a variation, caused by a slight change of surface, shows that there is something already going on at the metal-air surface to vary. I am not aware of anything that can be done to a metal-metal junction that will affect the Volta force—except of course heating it. Lord Kelvin, however, would say, I suppose, something like this:—In a simple Volta electrostatic arrangement there are three boundaries, zinc/copper, copper/æther, æther/zinc, and there is a contact-force at each. The Volta effect observed is the sum of the three. Its seat is mainly at the zinc-copper junction; it is due to the attraction of the two metals for each other, and its value is calculable approximately from their energy of combination. The effect of any contact-force there may be at the other boundaries is to modify the observed magnitude of this

* To this I should like to add that the gigantic differences of potential caused by the contact and separation of insulators, or of a metal and an insulator, *e. g.* the contact and separation of mercury or tin amalgam and glass, are very familiar; and the great change, sometimes even of sign, caused in this potential-difference by a modification of surface, for instance by using glass whose smooth or burnished surface has been ground with sand-paper, is well known, and inevitably suggests some cause, whatever it may be, akin to that which causes a less violent but somewhat similar drop of position in a volta series when a burnished surface of zinc is scratched with sand-paper.

† The following quotation from Lord Kelvin's Royal Institution Lecture may be made:—The result of the burnishing experiments "shows that the potential in zinc (uniform throughout the homogeneous interior) increases from the interior through the thin surface-layer of a portion of its surface affected by the crushing of the burnisher, more by ·32 volt than through any thin surface-layer of portions of its surface left as polished and scratched by glass-paper." If I may say so, I should express this fact in exactly the same way, viz. that the natural potential-difference between clean scratched zinc and dry air can be increased one third of a volt by burnishing its surface with a steel tool. (A remarkable fact and one not easy to explain). But I should not go on, as the next sentence goes on:—"The difference of potentials of copper and zinc across an interface of contact between them is only about $2\frac{1}{2}$ times the difference of potential thus proved to be produced between the homogeneous interior of the zinc and its free surface, by the burnishing."

true and chief Volta force. The Volta force also varies in a subordinate manner with temperature, and hence there is a minute reversible Peltier evolution of heat at the zinc-copper junction whenever a current passes. [*End of hypothetical quotation.*]

Size of Atoms.

The doctrine of Lord Kelvin about the energy of attraction of plates of molecular thinness approaching each other to within molecular distance and so virtually forming an alloy, though admirably ingenious, proves as it seems to me too much, *i.e.* gives too sharp an upper limit to molecular dimensions; for, as he shows with great clearness in a table in his Royal Institution discourse already often quoted, the estimate of 10^{-9} cm. for molecular thinness and distance would cause the combining metals to rise to a temperature of 7900° C., or more than the probable temperature of the sun! Yet there is nothing out of the way in a molecular estimate of 10^{-9}. On the other hand, an estimate of 10^{-8} yields a rise of temperature of 79°, which appears to agree very well with the order of magnitude (36° Roberts-Austen, 77° Dr. Galt) obtained in actual experiments on the heat of formation of a zinc-copper alloy, a truly remarkable coincidence. One, however, that would be spoilt if the molecular-dimensions' estimate were only *halved*, for the theoretical rise would then be over 300°.

With deference I submit that the air-films, with which the metals employed in ordinary voltaic experiments are certainly coated, ought to be left on them for the data to be applicable; and in that case, when the molecular approach is made, the metals are not so much alloyed as *burnt*; so that a rise of temperature of several hundred degrees is reasonable, and a corresponding molecular estimate of say half 10^{-8} is also reasonable and likely. But in the metallurgical making of brass it is not to be supposed that the metals entering into combination are coated with air-films of any importance, for their proportion of free surface is insignificantly small. Hence I think the Volta effect has nothing really to do with the making of brass.

To carry out the above combustion process, that is the voltaic approach of metallic foil of molecular thickness and its investing oxygen coat, it is not necessary to complicate matters by using *two* metals; and in the Phil. Mag. for May 1885, p. 364, I endeavoured to make a somewhat precise estimate of molecular magnitudes in this way, and even to discriminate between the molecular dimensions of different metals, since x comes out proportional to the square root of the atomic weight divided by the square root of the density and multiplied by the square root of the heat of oxidation.

Chapter III.

Statement of the present condition of the Controversy.

The opposing sides of the old controversy used to be called contact theorists and chemical theorists. If this were ever a correct designation it is not the correct designation now.

There is an E.M.F. at a junction of two media, and this may be called the E.M.F. of contact, without any view as to its origin or magnitude.

This contact E.M.F. may be caused in a chemical or in a physical manner. It is generally admitted, I believe, that the E.M.F.s concerned in thermoelectricity are caused in a physical manner. It is generally admitted, I believe (I think the statement will not be objected to in substance though its purposely crude wording may be criticised), that the E.M.F.s concerned in Volta contact electricity, as well as in ordinary voltaic electricity, are caused in a chemical manner. In saying this I am not intending any controversy, I am only trying to state. The best plan perhaps at this stage is to be more explicit.

I hold that Volta forces can be calculated, as regards their major part, from the differential energy of combination of the metals with oxygen (or other active surrounding element), and I have justified this, more or less completely, by numerical calculations (B. A. Report, 1884; Phil. Mag. April 1885, § 17, pp. 272, 274, &c.) based upon the great fundamental voltaic-cell paper of Lord Kelvin in 1851.

Lord Kelvin, on the other hand, holds that the Volta force can be calculated, as regards its major part, from the chemical affinity of metals for each other; or, to be specific, from the energy of combination of zinc with copper in the formation of brass. It is true that for the calculation to be made in this order, an estimate of molecular dimensions must be postulated, whereas his original calculation treated the order of calculation inversely; but this is of no immediate consequence, and no one at the present date has any serious compunction at including the order of ordinary molecular dimensions among data practically known.

Very well then, unless this statement be objected to, I say that the opposite camps are involved both of them in contact views, and both of them in chemical views. It is not a question of whether physical contact or some form of chemical action is operative; it is a question of which of several contacts is the really effective one, and what kind of chemical action or chemical affinity is the active cause.

Is it the contact and the chemical affinity across the metal-metal junctions, or is it the contact and chemical affinity across the metal-air junctions?

The opposite camps are thus metallic versus dielectric.

I say oxygen. Lord Kelvin scouts oxygen and says brass. He also says æther, and I am willing to say æther, too, in some sense later on, if the evidence permits; but meanwhile, it is more commonplace and safer to say oxygen: for undoubtedly oxygen is there.

Lord Kelvin does not deny the possibility, perhaps the probability, of *some* E.M.F. at the metal-air or metal-vacuum junction, which contributes to the total Volta effect, but I apprehend that he does not believe its contribution to be great. Nor, on the other hand, do I deny the existence of an E.M.F. at the metal-metal junction, which again contributes to the total observed Volta effect, but I assert that we know its contribution to be small.

How big it is in my opinion is answered above in Chapter I. It is measured by the Peltier effect (Profs. Ayrton & Perry will here ejaculate, Stuff and nonsense! and Lord Kelvin will support them. R. I. Discourse, §§ 23 & 24.) The metal-air force is of the order volts. The metal-metal force is of the order millivolts.

Not much is known for certain about the Peltier or

physical forces at metal-liquid or liquid-liquid junctions, and they are difficult to measure; but they have been partially measured by Bosscha and by Bouty (see reference in my Report, p. 189, Phil. Mag., March 1885). They may also be theoretically estimated in the way suggested by Dr. Hopkinson in the Phil. Mag. for October 1885. M. Bouty thinks that he has shown that the relation $\Pi = TdE/dt$ holds at such junctions. What is certain is that if all the Peltier forces are added algebraically together in a complete circuit at uniform temperature, then, whether the completion be dielectric or electrolytic, $\Sigma(\Pi) = TdE/dt$; as Lord Kelvin and von Helmholtz have proved, for the two cases of (1) dielectric medium, (2) electrolytic medium, respectively.

Thus the difference or controversy is no longer, if ever, a difference between contact and chemical action; a chemical contact seems to be admitted now by both sides, but there is an essential and by no means a nominal difference on the question of the localisation both of the E.M.F. and of the chemical contact. I locate the Volta force (the chief part of it) and the chemical contact too, at the air boundary of each metal. Anything occurring at the metallic junction I regard as physical. Lord Kelvin locates the chief part of the Volta force at what he is able to regard as an effectively chemical contact, the zinc-copper junction; and I presume that he regards whatever happens at the metal-æther boundary as physical.

If asked why I maintain the zinc-copper junction to be chemically inert, I pass a current across it, and point to the absence of any chemical change there. If I am chaffed for over-veneration for oxygen and its nascent or potential effect on metallic surfaces, I again pass a current, by whatever means are possible, and show oxide (or more likely sulphate, this acid radical having much the same differential energy of combination with zinc and copper as oxygen has. Nor is that of chlorine very different. But one can show oxide itself if pressed, by using a lead electrode, or a copper one with an intense current).

But if told that this cannot happen always, and that the energy of combination with something (say sulphur) must be very different, I fully admit it; but then I say that in an atmosphere of which the active ingredient is sulphur (say sulphuretted hydrogen) the Volta force will be different too. Mr. J. Brown * has shown that it is, but the conclusiveness of the experiment is evaded by pointing to films of sulphide which introduce fresh contact-forces and complicate the circuit.

So here begins a conflict of experimental testimony and interpretation. Lord Kelvin depends on the experiments of Mr. Bottomley in a hydrogen vacuum †. I am now able to appeal to the still more elaborate and to me satisfactory experiments of Mr. Spiers (Phil. Mag. Jan. 1900). In so far as they are not conclusive, and I must assume that they are not yet rigorously conclusive, they should be repeated and improved. The research is bound to be laborious and troublesome if properly executed. I will only say, in case any young and energetic experimenter takes the matter up, that Mr. Spiers seems to me in method and motive extremely on the right track.

The question of Expression in terms of Potential.

Superposed upon the controversy as to the junction at which the main contact force really exists, there has arisen another controversy, chiefly concerned with words or modes of expression, concerning the most convenient method of defining the potential of a conductor, and especially the difference of potential between two metals.

It has been customary with certain writers, among them Professor Perry, to define the potential of a conductor as the potential of a point in the air near it,—the work that must be done to convey unit electric charge from the earth (or any place of zero potential) to a point very near the conductor; or, better, to a point in a hollow surrounded by the conductor.

* Phil. Mag. August 1878 *et seq.*

† Mr. Bottomley sums up thus (British Association Report, 1885, pp. 901–903):—"The result of my investigation, so far as it has gone, is that the Volta contact-effect, so long as the plates are clean, is exactly the same in common air, in a high vacuum, in hydrogen at small and full pressure, and in oxygen. My apparatus, and the method of working during these experiments, was so sensitive, that I should certainly have detected a variation of 1 per cent. in the value of the Volta contact-effect, if such a variation had presented itself."

I have never been able to see why these philosophers hesitate to complete the process and give the charge actually to the conductor. Why do they stop short outside? The completion of the process is so easy that precaution is needed to avoid it.

It is usually an easier matter to measure experimentally the potential of a conductor, a thing which can be connected directly by a wire to an electrometer, than it is to measure the potential of a point in air or other insulator; the latter operation requiring a water-dropper or a smoking match or an infinitely sharp point.

It has been authoritatively stated that the heterogeneous molecular structure of a metal makes it difficult to affix any meaning to the potential of an inaccessible point inside it; but if this difficulty were a real one it would surely apply still more forcibly to a heterogeneous molecular insulator. The difficulty, however, seems to me purely imaginary from any practical point of view; we need have nothing to do with the molecular structure of a metal in this connexion; by hypothesis it is a conductor, its molecules are all effectively connected, and the potential of one point is the potential of all. Nothing can be simpler than the statement that in electrostatics every part of a conductor is at one potential. We need not put on what Lord Kelvin has somewhere called a molecular microscopic binocular, and examine the atoms on all occasions.

If asked to *define* the potential of a metal, with reference to some arbitrary standard conductor given as zero, I have only to say—the work needed to transfer a small unit charge from one to the other: or the potential energy of each unit of charge existing on the metal, provided the removal of that charge does not appreciably affect the potential: or, better, dW/dQ. Where is the difficulty? The process of charging a conductor is experimentally easy, why not introduce the process into the definition *.

* Nevertheless, Lord Kelvin says:—"There has been much of wordy warfare regarding potential in a metal, but none of the combatants has ever told us what he means by the expression. In fact, the only definition of electric potential hitherto given has been for vacuum or air or other fluid insulator. Conceivable molecular theories of electricity within

If the potential of a metal is high, positive electricity tends to flow away from it or out of it, and does flow if a path free from obstruction is provided. If the potential of a metal is low, positive electricity tends to flow into it.

That the potential of an isolated piece of zinc is lower than the potential of an isolated piece of copper is proved (I hold) by the undoubted fact that a rush of electricity takes place from one to the other directly they are put into contact. The electricity is driven by the difference of potential: the metals may be brought near and facing each other so as to constitute the plates of a condenser of appreciable capacity; then let them touch, and the flow automatically occurs. If potential is that which determines electrical transfer, if it be the property whose gradient is electromotive intensity, the natural statement about the previous potential of the two metals is obvious. Before contact they were at different potentials; while they are in contact they are at the same potential. The difference of potentials is due to the unequal atomic forces straining at their frontier; the equality of potentials is caused by the equalising electric transfer, *i. e.* to the production of charges which hold the still existing chemical forces in equilibrium. What can be simpler than this mode of expression?

The other mode of expression says that the metals in contact are at different potentials; meaning that the air near each is at a different potential. If that is the meaning, why not say so? It is an undoubted fact. The step of potential in the air is experimentally demonstrable, the only question concerns the step of potential across the junction of the two metals.

To transfer a unit charge from copper to zinc in contact involves no appreciable work, whether it goes through the

---

a solid or liquid conductor might admit the term potential at a point in the interior; but the function so called would vary excessively in inter-molecular space," &c. (Footnote to R. I. discourse, p. 13.)

He goes on, however, "It would also vary intensely from point to point in the ether or air outside the metal at distances from the frontier small or moderate in comparison with the distance from molecule to molecule in the metal." And with this sentence I can quite agree; as well as with the bare fact stated at the end of the previous sentence.

junction or whether it goes out and through the air*. To transfer electricity from *near* the copper to near the zinc involves a definite amount of work, provided that by "near" we do not mean within molecular range; and it requires the same amount of work whether we carry it through the dielectric, or whether we plunge it into the copper and then pull it out of the zinc. If this statement of fact be not disagreed with, it is surely more convenient to say that the zinc and copper in contact are at the same potential, and that the air or space in their immediate but not molecular neighbourhood differs in potential †. In other words, that there is a

* I think this is the clearest form of statement. It is possible to say there is a cyclical E.M.F. round a circuit like this with two different paths, and that the resultant force is opposed by air resistance; but when the gradient of potential in the air is taken into account as part of the E.M.F., there is equilibrium, and the cyclical E.M.F. is zero. It is merely a question as to which is the most convenient form of expression; there is no controversy in it: the question is touched on again at the end of the next footnote. All that I require for my purpose is the admission that no appreciable work is needed to transfer electricity across the copper-zinc junction; while work is needed to transfer it in the air from near the copper to near the zinc.

† It may be permissible to quote a sentence from Dr. John Hopkinson's contribution to the discussion on the same subject at the Institution of Electrical Engineers (Journal, 1885, p. 235) :—"Professor Perry defines the difference of potential between a piece of copper and a piece of zinc to be the energy required to transfer a unit of electricity from near the copper to near the zinc through the dielectric. On the other hand, as I understand the matter, Dr. Lodge, and probably Clerk Maxwell, define the difference of potential to be the energy required to transfer a current of electricity from the copper to the zinc through the junction between the copper and the zinc. Now, it appears to me that Dr. Lodge has very clearly shown that his way of looking at the matter is the more convenient; for, in the first place, it expresses, with the greatest ease, all facts that have been observed. When copper and zinc are in contact with air there is a difference of potential at the three junctions, and the sum of the three differences of potential is not equal to zero; and we have no difficulty in explaining all the electrostatic effects that we have obtained. Dr. Lodge has so clearly proved the convenience of his way of looking at the matter that it would be only wasting your time if I were to attempt to reproduce his arguments. It seems to me better to take that basis and pursue it. Now, if we take that basis as our starting-point, it is very easy to set forth all the theory of thermoelectricity, and the formulæ given at the end of Clerk Maxwell's chapter on that subject can be proved with the greatest ease; it is also easy to extend these formulæ to the case in

sudden step of potential in crossing the frontier from metal to air, a step which I have reckoned as about 1·8 volts up out of zinc, and about ·8 volt up out of copper, and so a (roughly speaking) 1 volt gradual descent of potential in the air or space near them.

I have found in the reported discussion at the Electrical Engineers (Journ. Inst. E. E., 1885, vol. xiv. p. 233) the following ingenious and evidently careful note added later by Prof. Perry. I will first quote it, and then paraphrase it into what I consider greater simplicity of form :—

"To be quite correct, let what I have called 'tension' in a metal a, at the absolute temperature t, be denoted by ${}_tP_a$; and the increase of potential from a to a metal b, as measured inductively, by ${}_tV_{ab}$, then the electromotive force of a thermo-electric circuit of two metals, a and b, whose junctions are at temperatures t_1 and t_2, is

$$E = {}_{t_2}V_{ab} - {}_{t_1}V_{ab}.$$

The Peltier effect at a junction is

$${}_t\Pi_{ab} = {}_tP_b - {}_tP_a - {}_tV_{ab},$$

where

$${}_tV_{ab} = (k_b - k_a)t\left\{T_{ab} - \tfrac{1}{2}t\right\} + C_b - C_a,$$

$${}_tP_a = \tfrac{1}{2}k_a t^2 + C_a.$$

k_a and C_a are constants peculiar to the metal a, and T_{ab} is a constant peculiar to the two metals a and b.

Hence

$${}_t\Pi_{ab} = t \cdot \frac{d}{dt} \cdot {}_tV_{ab},$$

or the Peltier effect is proportional to the absolute temperature and to the rate of change of the contact-force with temperature."

So far the quotation ; now for its interpretation.

---

which we have chemical action going on, and to foreshadow the precise method by which we may be able to ascertain exactly what the difference of potential is between copper or zinc or other metals and electrolytes."

He then makes some further remarks of considerable interest which were expanded by him later in the Phil. Mag. for October 1885. All I would add to the above admirable statement is that the emphasis required is upon the word "near" in the 3rd & 4th lines, and that the immediately following specification of path—"through the dielectric"—is not really essential. The potential-difference is independent of path in this case. Replacement of dielectric by electrolyte, unable to stand the potential-gradient statically, may be said to destroy equilibrium and yield a cyclical E.M.F. and a current ; but, strictly speaking, the cyclical potential-difference is still zero, when resistance to steady current is taken into account; always zero, in fact, even within the variable stage, when acceleration-reaction or self-induction is likewise included.

PARAPHRASE :—Let the step of potential from inner metal to outer air across the boundary film be p, and let it be a function of temperature such that

$$p=m+\tfrac{1}{2}kt^2. \quad \text{[Assumption.]}$$

Let the change of potential from air near metal a to air near metal b be V_{ab} or V; it follows that this also will be a function of the temperature. Because, if Π be the E.M.F. located at a junction of the two metals, the total change of potential on a journey all round from a to b and back, out *via* the air and back *via* the junction, is zero; or

$$p_a+V_{ab}-p_b+\Pi_{ba}=0.$$

[This is only a roundabout way of saying that the Volta effect is the sum of the three junction-forces.]

Further, if we construct a complete circuit of the two metals and keep their junctions at t_1 and t_2 respectively, and then travel right round outside the circuit, keeping in the air near the metals, we shall encounter an E.M.F.,

$$E=V_1-V_2.$$

[Here again there is an assumption :— viz. that the Thomson force in a metal bar, with its ends kept unequally hot, will show itself outside not as a slight gradient of potential in the air along and near the rod, but as a slight modification by temperature of the step of potential in its boundary film. This assumption is therefore quite consistent with the one made just above, $p=m+\frac{1}{2}kt^2$.]

Further, let us assume that $\Pi=(k_a-k_b)t(t_0-t)$; then it follows by common algebra that

$$\Pi/t=dV/dt.$$

So with all these assumptions, and all this needless and artificial attention to external occurrences, nothing is obtained beyond the admitted laws of the closed thermoelectric circuit, most simply written thus :—

$$\Pi=tf'(t); \qquad E=f(t_1)-f(t_2).$$

See, for instance, my footnote to p. 270 of Phil. Mag. for March 1886, or see pp. 265, 266 of the same paper.

I trust that I have made it clear that I do not regard this matter of the definition of potential as of the same importance as the seat of E.M.F. in the pile. In no other part of this paper have I paid any attention to the wording of propositions in terms of potential; it is a subsidiary dispute superposed on the main disagreement, which latter is in no sense a matter of words or definition.

Nevertheless the two disputes are somewhat closely allied, and the clarification of one ought to result in pacification of the other.

Chapter IV.

My advocated mode of regarding the Mechanism of the Chemical Contact Force.

I have elsewhere tried to explain the mode in which I suppose the oxygen to act to bring about that momentary transfer of electricity which occurs across the junction of two metals the instant they are put into contact. The facts postulated are (1) the fact of chemical affinity between oxygen and metal, of amount different for different metals, and (2) the fact that oxygen atoms are, at any rate some of them, negatively charged.

On the strength of these facts I see a surrounding layer of straining oxygen atoms seeking to move up to the zinc, but unable to do so because of "the impossibility of conferring an absolute charge on matter," as Faraday in his ice-pail experiment put it, or because of the "incompressibility of electricity," or because there is no avenue for the supply of electricity of opposite sign. Consequently we have a state of siege, a kind of incipient polarisation, but no charge. The zinc is somewhat in the condition of an insulated sphere surrounded by a concentric negatively charged shell, so far as the interior of the shell is concerned : outside the conditions are different. And the cause of the stress is different : there are no lines of electric force, there is no *electrical* difference of potential; whatever force or potential-difference exists between isolated zinc and air is chemical in its nature—the result of chemical affinity. The force may be due, as Helmholtz puts it, to a specific attraction of zinc for electricity, or it may be due to a specific attraction for oxygen; but to whatever it is due, so long as the surface is homogeneous, it is inoperative, it results in no energy transformation or work done, it is a deadlock.

Now establish connexion at one point with some neutral substance, say platinum, or a less strained-at substance, like copper ; instantly an avenue for the relief of the strain is provided, positive electricity flows across the junction into the zinc, takes up its abode on the surface facing the oxygen atoms, the oxygen atoms approach slightly nearer all round, the surrounding molecules all round are polarised, their negative poles inwards, their positive poles outwards, a double layer is set up on the zinc, and lines of force appear all through the surrounding dielectric. These lines reach from the zinc free-surface to the copper free-surface, distributing themselves according to the ordinary rules of electrostatic capacity, and their terminations on the copper constitute its negative charge. In other words, the oxygen atoms are slightly further removed than before from the copper, and it is to that extent protected even from its previous small tendency to oxidation; on the other hand, the oxygen atoms have approached slightly nearer to the zinc. Thus my doctrine is not a doctrine of chemical combination, but of chemical approach. The oxygen atoms approach nearer the zinc, recede further from the copper, and by these slight motions produce the whole Volta effect.

But this it may be said is nothing but a voltaic cell, except that the electrolyte is replaced by a dielectric, and so the current is momentary, not continuous. Precisely, that is my meaning.

But then, it will be objected, if a current did pass, there would be actual oxidation; since for every electric unit that passes in a cell there must be an electrochemical equivalent of zinc oxidised. To this I agree.

Well then, pursues the objector, consider a condenser of great capacity with its alternate plates made of zinc and copper, very close together, and then short-circuit it at some point; there will instantly be a great flow of electricity across the junction to charge the condenser, and accordingly there will be a perceptible amount of zinc oxidised. Your mere approach of oxygen atoms without combination will no longer serve.

At this point I appeal to arithmetic. The charge on an atom is of the order 10^{-11} electrostatic units. The oxygen atoms immediately in so-called contact with the zinc are constituents of its condensed air-film, a film so dense as to be virtually a liquid, since it is subject to the intense molecular pressure. Not really a liquid, because it is far above the critical point, but not perceptibly differing from a liquid in density. If this be objected to, it is not essential to my argument, because there is plenty of margin; only it seems to me likely to correspond roughly with the truth. The number of atoms per square centimetre in the layer facing the zinc will therefore be of the order 10^{16}. Of these only a portion will be oxygen; and probably it may be said that only a small proportion are ionised and so open to combination. But as I do not need combination, I see no necessity to postulate ionisation, or to care what proportion are in this condition. Certainly I need molecules capable of being polarised electrically, but that at the present date will hardly be denied to any material molecules whatever.

Very well then, if for any reason polarisation takes place, the surface-density of the charge on either face of the double layer can be $10^{16} \times 10^{-11} = \sigma = 10^5$ electrostatic units per square centimetre, at a maximum. The mechanical tension across the layer is $2\pi\sigma^2$, and the intensity of electric field there is $4\pi\sigma$, which is also the gradient of electric potential.

Now the thickness of the layer is of molecular dimensions, 10^{-8} centim. say; hence the difference of potential between its faces, that is the difference of potential between the zinc and the air in its neighbourhood, is $4\pi\sigma z$, which is of the order $(12 \times 10^5 \times 10^{-8} = 10^{-2})$, one-hundredth of an electrostatic unit, or 3 volts. Now plainly this is of the right order of magnitude. The Volta effect observed *can* be explained by such an electrical double layer, by such a chemical layer of straining oxygen atoms, at a metallic surface *.

That it can be equally well explained for metals in absolute vacuum or free æther I would not presume to deny; but I cannot do the arithmetic for that case, because of an utter lack of data. It is to be observed, however, that the only gaseous or material substance needed (other than metal) is the coherent film on the surface; all the rest of the gas is merely a dielectric medium for the transmission of electrostatic induction or lines of force, and for this function an absolute vacuum serves perfectly. Consequently, merely removing the plates into the receiver of an air-pump will make no practical difference.

But still, so far, I have not met the hypothetically suggested difficulty about the condenser of large capacity and the amount

* See also a footnote to p. 379 of a paper in the Phil. Mag. for October 1885, where I reckoned practically the same thing.

of real chemical action needed to account for the current required to charge it to the voltaic difference of potential.

We have only to proceed with the calculation.

The condenser formed by the opposing surfaces of zinc and copper is a condenser of the same (or comparable) area to that molecular condenser just postulated between either metal and the air. The only difference is that these last are of molecular thinness, whereas the condenser formed by well-fitting plates of zinc and copper is of a moderate thinness, z, say the hundredth part of a millimetre, or something of that order.

Hence, when the plates are approached from infinity to this distance z, and are put into contact to establish the state of polarisation, there will be a rush of charge out of the molecular condensers into the newly formed condenser; for they are condensers in series, and every unit gained by one must be lost by the other.

But since the molecular condensers are to retain their original potential-difference, this supply of electricity can only be given by a change in capacity, *i. e.* a change in their molecular thickness x, such that

$$-\frac{dx}{x^2}=\frac{1}{\infty}-\frac{1}{z},$$

or

$$dx=\frac{x^2}{z}=\frac{10^{-16}}{10^{-3}}=10^{-13}\text{ cm.},$$

the hundred-thousandth part of molecular magnitude.

Thus the diminution in thickness of the superficial double layer, or the approach of the oxygen atoms, in an extreme case, has only to be a hundred-thousandth part of their whole distance. This can hardly be called chemical combination: indeed it is a scarcely perceptible chemical *approach*.

Put in another way it amounts to this: that if the oxygen atoms varied their distance from the zinc and the copper respectively (one decreasing, the other increasing) by so much as a tenth of 1 per cent., the electricity so set free or transferred from one surface to the other would be able to charge an ordinary condenser of the same area in the intervening space to a density of 100 C.G.S.; which is considerably more than ordinary air can stand.

The only way to get a perceptible amount of real oxidation, that is an approach over the whole molecular distance, is either to bring the condenser plates themselves within molecular distance of each other, which would squeeze out the films and modify the whole action, or else to bring them within molecular distance of different parts of one and the same liquid conductor (so that the whole strain is thrown upon the molecular film-condensers and none upon any intervening material); and this is exactly what is done in an ordinary voltaic cell.

CHAPTER V.

Osmotic Pressure or Diffusion View of a Voltaic Cell.

The osmotic pressure hypothesis, strictly speaking, perhaps hardly belongs to my present subject, since whatever controversy or question has gathered round it is of a different kind from that concerning the seat of the E.M.F. The osmotic pressure hypothesis does not concern itself apparently with the Volta electrostatic effect, but with the mechanism of propulsion in a voltaic cell: it hardly cares to discriminate between the two cases, for they appear in many respects much the same; and it locates the seat of the E.M.F. quite definitely at a junction on one side of which at least there is a liquid, or substance with loco-mobile atoms, and not at a metal-metal junction: it deals entirely with moving ions, such as cannot exist in true solids. Moreover it is not a hypothesis really distinct from that of the recognized chemical combination associated with a current-producing cell, only it regards that chemical combination from a different point of view—from the dissociation or ionisation point of view. Instead of saying that zinc is oxidised or that zinc sulphate is formed, it professes to regard atoms of zinc as shot off into the liquid in a non-combined or free form: ultimately, no doubt, to enter into real combination, but existing for a time, and always in a certain proportion, in the form of practically free ions. Thus the full discussion of the osmotic pressure hypothesis is essentially chemical in its nature, and need not here concern us. At the same time so many people are interested just now in this remarkable view of chemical combination, that I refrain from passing it by altogether; though, as it is not a hypothesis specially and long familiar to me, my remarks on it will not have any particular weight. Fortunately there are several writings to which a student of such matters can be referred—the treatise of van t' Hoff, the work of Arrhenius and of Planck, and the textbooks of Ostwald and of Nernst, especially of Nernst; and likewise the excellent Report on these subjects presented to the British Association in 1897 at Toronto by Mr. W. C. D. Whetham, as well as to his book on 'Solution.' Furthermore, I may refer to the works of Dr. J. Larmor, especially Phil. Trans. vol. cxc. p. 270; and to a paper by Prof. Poynting in Phil. Mag. vol. xlii. p. 289 (1896), where it is shown that the dissociation hypothesis need not be pressed too literally, that solution may be a kind of chemical combination, and yet that all the facts deduced on the free ionisation hypothesis can be true. This compromise is, however, criticised by Dr. Larmor in a footnote to page 273, *op. cit.*; and Boltzmann is able to regard osmotic pressure as due to a true molecular bombardment, seeing that in the liquid state the molecular velocities should be the same as in the gaseous state at the same temperature, the mean free paths alone being smaller. On the other hand, it has been frequently claimed that any extension of gaseous laws, *i. e.* of laws deduced from the collisions of free particles, to the extremely hampered condition of a liquid, where the particles are always within each other's molecular range, must be illegitimate.

Without attempting to decide these moot points, we will here consider the question open whether the statement in terms of free ions is only a mode of expression, or whether it corresponds with the real facts; corresponds with them, that is to say, not only in result, but in actual detail. Prof. Poynting's paper, for instance, does not decide against the latter possibility, it only attempts to show that it is not really *necessary* to grant free ions as a physical reality; but Mr. Whetham evidently inclines to the idea that they are at least free of each other, and adduces an argument in favour of that view (pp. 233, 244, B. A. Report 1897) which is worthy of attention. But, while assuming that opposite ions are free of each other, he considers that they are both attached to the solvent, and are by no means really and bombardingly free, as are the molecules of gases. Prof. Poynting appears to agree *; but, as said above, the agreement is not universal.

The most successful application of Nernst's theory is to the calculation of the feeble E.M.F. of cells consisting of one metal but two liquids, especially when the two liquids differ only in concentration of one and the same salt.

The ideas underlying all such calculations are (*a*) that the ions migrate for two reasons: one their own random motion whereby they naturally diffuse on the average from the more concentrated to the less concentrated solution, carrying their charges with them, the other an organized motion caused by the propulsive influence on these same charges of the gradient of potential caused by their gradual diffusive accumulation; and (*b*) that in the steady state ultimately attained the rate of migration of opposite ions must become equal: otherwise the liquid would be constantly altering its composition in places.

Indeed the fact that equivalent quantities of anions and cations are liberated at the electrodes, combined with the definiteness of ionic charge, shows that in electrolysis the resultant effective velocities of anion and cation substance relatively to the electrodes are equal †.

Consider now the charged atoms or ions in a cylindrical space of volume $A\,dx$, with n of them in each unit of volume.

Let them be subject to an applied potential gradient dV/dx, and also to a self-excited osmotic pressure gradient dp/dx; then if e is the charge on each atom, the electrical force it experiences is

$$e\frac{dV}{dx},$$

while the mechanical or osmotic force it experiences is the

* Nature, 1896, vol. liv. p. 571; vol. lv. pp. 33, 78, 150, &c.

† This statement seems to require a word of explanation or expansion. So much stress has been laid upon the fact of different intrinsic velocities of opposite ions when under the same conditions, that it is apt to be forgotten that in ordinary electrolysis the conditions are such as to force equal, or at least electrically equivalent velocities, on anion and cation substance taken as a whole, in the steady state. The ions are propelled not by outside forces acting from the electrodes, but by internal actions: the readjustment of the whole material must be attended to. The intrinsic speed of a projectile is greater than that of the gun-recoil, yet, after a battery has fired, the centre of gravity of the whole is no nearer the

whole pressure difference on the element shared among the total number of atoms, which is

$$\frac{\mathrm{A}dp}{n\mathrm{A}dx} \quad \text{or} \quad \frac{1}{n}\frac{dp}{dx}.$$

This latter mechanical force acts equally on anions and cations, but the electrical force acts oppositely on each set; so the total force on the one set is the sum, on the other set is the difference, of these two forces. If u_1 and v_1 are the usual migration constants representing the velocities of anion and cation respectively under unit potential gradient, their velocities under unit force will be u_1/e and v_1/e, and the actual velocities under the above forces can easily be written down. But the actual velocities in the steady state will be equal; therefore, as the condition of steadiness,

$$\frac{u_1}{e}\left(\frac{1}{n}\frac{dp}{dx}+e\frac{d\mathrm{V}}{dx}\right)=\frac{v_1}{e}\left(\frac{1}{n}\frac{dp}{dx}-e\frac{d\mathrm{V}}{dx}\right)$$

which simplifies to

$$ne\frac{d\mathrm{V}}{dx}=\frac{v_1-u_1}{v_1+u_1}\frac{dp}{dx}.$$

At this stage it is customary to introduce the osmotic pressure gas-analogy, writing the characteristic equation of a mass M and volume V,

$$p\mathrm{V}=\mathrm{MRT},$$

where R is a characteristic constant, viz. the specific heat of expansion per gramme of perfect gas, which for hydrogen is accidentally almost equal to J, the specific heat of water.

---

target than before. The mass-speeds are equal and opposite. In the electrical case the charge-speeds are equal and opposite. Direct experiments on specific migration velocity (such as those I described in the Brit. Assoc. Report for 1886, pp. 393, 408) are all made in purposely heterogeneous solutions; so that the motion of specific portions of matter can be discriminated and followed. Even in such heterogeneous liquids the current-strength is the same across every section; but the local potential-gradient depends on the local conductivity, and the concentration is arbitrary: it is still true, however, that the potential-gradient is the same for anion as for cation at any given place; and hence if they do ultimately travel at the same pace in spite of intrinsically different ease of travel, the fact must be expressible as due to an accumulated osmotic pressure-gradient which opposes the quicker and helps the slower.

Writing $\frac{M}{V}=\rho=nm$, and considering the gas pressure p to be the osmotic pressure,

$$dp=m\mathrm{RT}\,dn,$$

at constant temperature.

And so finally the steady difference of potential due to diffusion between two solutions which differ only in concentration, one containing n the other n' active atoms to the litre, is

$$\mathrm{V}-\mathrm{V}'=\frac{m}{e}\,\mathrm{RT}\,\frac{v_1-u_1}{v_1+u_1}\log\frac{n}{n'}.$$

I have written this out fully because it is usually somewhat slurred over, the electrochemical equivalent m/e omitted, the argument rendered obscure, and the dimensions wrong.

It follows that although usually the E.M.F. thus set up is very small, yet by having one of the solutions extremely weak (n' nearly $=0$) the E.M.F. generated may be made surprisingly high, that is a large fraction of a volt; provided the anions and cations concerned do not diffuse at nearly the same rate under the same circumstances.

For instance, if silver is placed in a liquid containing some chloride, no molecules of silver can accumulate in any quantity in the solution, and accordingly the concentration will be almost infinitesimal, and the E.M.F. comparatively high; thus Ostwald gives the following as a cell which has, and ought on his calculation theoretically to have, an E.M.F. of ·51 or ·52 volt.

$$\mathrm{Ag} \mid 0{\cdot}1\ \mathrm{AgNO_3} \mid \mathrm{KNO_3} \mid 1{\cdot}0\ \mathrm{KCl} \mid \mathrm{AgCl} \mid \mathrm{Ag},$$

the KNO_3 being introduced for obvious reasons to prevent precipitation of adjacent solutions, but being without influence on the resultant E.M.F.

To calculate the total E.M.F. of a cell, however, it is necessary to take the other boundaries of the liquid into account. It is useless to complete the circuit by liquid, it must be completed by metals, and in the cases now being considered by one and the same metal. Each metal acts as a wall preventing effective diffusion of one at least of the liquid ions: the cathode prevents the cation from advancing, the anode prevents the anion; whatever else the metals may do will be the same at each end, and so balance. This is I believe the form given by Mr. Whetham to Prof. Nernst's argument.

Hence the total E.M.F of such a cell will be the sum of the three junction-E.M.F.'s, metal/liquid, liquid/liquid, liquid/metal :

$$E = R\epsilon T\left(\frac{u_1 - o}{u_1 + o}\log n + \frac{u_1 - v_1}{u_1 + v_1}\log\frac{n'}{n} + \frac{o - v_1}{o + v_1}\log n'\right)$$

$$= \frac{m}{e}RT \cdot \frac{2v_1}{u_1 + v_1}\log\frac{n}{n'}.$$

There is here, fortunately, no question about an innocent-looking but controversial metal/metal junction.

The remarkable thing about this equation is that the data on the right-hand side are all known, and that therefore E can be calculated in absolute measure.

Thus since the characteristic gas constant R, the absolute specific heat of expansion per gram of perfect gas, varies inversely with m the atomic weight, it follows that Rm is an absolute constant, the same for every substance, or at least only differing by a simple multiple, having twice the value for electrolytic as for undissociated molecular substances ; hence we may take its value as double that for, say, hydrogen. For hydrogen $m/e = \cdot 00010352$ gramme per coulomb, and $R = 41\cdot 6 \times 10^6$ ergs per gramme degree.

Hence at 15° C.

$$E = 2 \times 4\cdot 16 \times \cdot 00010352 \times 288 \frac{2v}{u+v}\log_e\frac{n}{n'},\text{ joules per coulomb}$$

$$= \frac{\cdot 114}{1+r}\log_{10}\frac{n}{n'},\text{ volts,}$$

where r is written for the Hittorfian migration ratio of anion to cation intrinsic velocity u_1/v_1. It is remarkable that nothing is wanted except this migration ratio and the concentration ratio n/n' to calculate in absolute measure the E.M.F. of such a cell, with the same metal at either end.

A table of correspondences between theory and experiment is quoted from Ostwald and Nernst by Whetham in his B. A. Report, and they agree very remarkably indeed when one considers that the numbers are absolute numbers and might have been in error by large quantities.

To illustrate this take a cell with decinormal solution of HCl on one side and centinormal solution on the other, so that the concentration ratio $n/n'=10$:—

For the migration ratio we might take Hittorf's own numbers (which he tabulates in the form $u/(u+v)$ &c.) ; but several more recent and presumably more exact determinations have been made, and it may be convenient to quote here the table of absolute velocities given by Mr. Whetham as embodying the most recent measurements in 1897. As I write them they give the ionic velocity, in mikroms per second, for a potential gradient of 1 volt per centimetre (a mikrom being 10^{-6} metre or 10^{-4} cm., about double the wave-length of green light).

Intrinsic Ionic Velocities.

	u_1.		v_1.
Cl	6·9	K	6·6
I	6·9	Na	4·5
NO_3	6·4	Li	3·6
OH	18·2	H	32·0
$C_2H_3O_2$	3·6	NH_4	6·6
$C_3H_5O_2$	3·3	Ag	5·7

So for HCl the migration ratio is

$$r=\frac{69}{320}=\cdot 216,$$

and hence the theoretical E.M.F. of such a cell as that just above spoken of, with a weak solution of HCl opposed to one ten times weaker and connected by any single metal, is

$$\frac{\cdot 114}{1\cdot 216}\log_{10}10=\cdot 0938 \text{ volt},$$

whereas Nernst gives it as observed at ·0926 volt. A very remarkable agreement when we consider how the above absolute constant $4RTm/e$, which I have reduced on the preceding page to ·114 volt, is constituted.

As another example take a cell containing two strengths of NH_4Cl solution, with the same concentration ratio 10 : 1. In this case

$$r=\frac{69}{66}=1\cdot 045.$$

So
$$E=\frac{\cdot 114}{2\cdot 045}=\cdot 0557 \text{ volt},$$

and Nernst gives it as observed at ·0546.

Once more, take a cell filled with KOH solution 10 : 1;

$$r=\frac{182}{66}=2\cdot 76,$$

therefore
$$E=\frac{\cdot 114}{3\cdot 76}=\cdot 0303 \text{ volt},$$

and it is observed at ·0348.

A cell of LiCl gives another example, strength as before decinormal and centinormal, that is ·1 gramme equivalent per litre on one side, and ·01 ditto on the other:

$$r=\frac{69}{36}=1\cdot 925.$$

So
$$E=\frac{\cdot 114}{2\cdot 925}=\cdot 0390 \text{ volt},$$

observed at ·0354.

Here more old-fashioned values for the migration numbers give a result more nearly agreeing with observation.

Another cell where the agreement is not so close is one of caustic soda, NaOH. Nernst quotes one with strengths ·235 gramme equivalents per litre on one side, and ·030 on the other, so that the concentration ratio is 7·83, while the migration ratio is $\frac{182}{45}=4\cdot 05$.

The observed E.M.F. was ·0178 volt, whereas the calculated is

$$\frac{\cdot 114}{5\cdot 05}\log 7\cdot 83= \cdot 0201 \text{ volt}.$$

At least that is what I get, but, as in several other cases, the column of "calculated" given by Nernst does not contain the same numbers as those here obtained. Sometimes they agree with observation better, but sometimes worse.

These agreements, though not perfect, are clearly far and away beyond anything attributable to chance, and they practically substantiate the essential features of Nernst's theory, a theory which seems to me to rank high among rational calculations of this kind. The success of this brilliant concentration theory of batteries with one metal and two liquids, or rather one liquid with different strengths, lends weight to the supplementary suggestions of the same great chemist*, that the E.M.F. of batteries containing one liquid and two metals can likewise be calculated on similar lines. So long as the metals are the same at either side, and only the solutions have to be considered, this diffusion theory clearly represents a great step in the direction of the truth, even if it cannot be asserted at present to be the whole truth.

But when the so-called solution-pressure of a metal itself has to be taken into consideration, the ground becomes less secure and satisfactory.

When zinc dissolves in an acid this hypothesis assumes not that the zinc is eaten away by combination with the acid radical, but that it as it were evaporates into the liquid, giving rise to free zinc ions having a certain osmotic pressure.

If asked why zinc should evaporate any more freely into liquid than into air, it may be answered (as Mr. F. H. Neville suggests), because the great cohesive force, Laplace's K, would resist evaporation into air, but would be far less effective in resisting evaporation into a liquid where the change of density is much less sharply emphasised at the boundary. Hence though a certain repugnance may naturally at first be felt to this evaporation theory, it may turn out a reasonable mode of expressing facts, and it demands careful consideration.

The above mere cohesion consideration would tend, however, to make all liquids too much alike in their solvent influence on metals; the form given to the theory by Mr. Whetham and Prof. Poynting, that the ions though free of each other are chemically attached to the solvent, seems far more plausible, and overcomes many difficulties. On either view the following statements hold good.

* Meaning either Ostwald or Nernst. I cannot pretend to discriminate the portion belonging to each, and I hope it is not necessary. Both have done admirable work in connexion with this subject. I find that the first theory of concentration batteries, based upon thermodynamic and vapour-tension considerations of a less simple kind than those here quoted, and not depending on any ionisation hypothesis, was given by von Helmholtz in 1878 (Wied. *Ann.* vol. iii. p. 201).

When a metal dissolves in a liquid there is no interchange of ions; cations pass into the liquid, but no anions necessarily pass out of it. Sometimes there is an interchange of cations—one set going in and another set coming out—as when iron displaces copper, or zinc displaces hydrogen. Probably some impurity in the metal, enabling galvanic action to occur, is essential to this process, but that is here not a matter of primary importance.

Consider now a Daniell cell of the simplest theoretical kind:

$$Cu \mid CuSO_4 \mid ZnSO_4 \mid Zn.$$

As soon as the circuit is completed, zinc goes into solution at one end and copper comes out at the other. The difference between the energy of combination of Zn and Cu with SO_4 gives the whole E.M.F.; since, as is well known, the dE/dt and therefore the $\Sigma(\Pi)$ happen in this case to be practically zero.

But what about the location of the E.M.F.? Leaving on one side the extraordinary contention in favour of the zinc-copper junction, there are three other junctions at which the chief portion of the E.M.F. might be located; and at first sight the liquid/liquid junction at the porous partition seems the most likely. For here it is that the actual exchange of Zn for Cu takes place; and hence it may be said, either heat must here be locally developed or else the energy must be used in propelling the electric current and developing the same quantity of heat elsewhere.

But, as Hopkinson virtually pointed out (p. 341, Phil. Mag. Oct. 1885), this involves the tacit assumption that the metallic ions do all the travelling, while the SO_4 is stationary;—in that case the metallic zinc which goes into solution would at first be merely passed on, and the real energy-production would occur where zinc atoms take the place of copper atoms, viz., at the junction of solutions in the porous partition. After that there is a mere passing on again, though now of copper.

But the migration ratio of the ions is in reality different from this, and the SO_4 is by no means really stationary. As another extreme assumption, suppose that the SO_4 ion did all the travelling; then the zinc sulphate solution is strengthened and the copper sulphate solution is impoverished, and the whole positive energy of combination is at the surface of the zinc, while at the surface of the copper there must be a smaller opposition force corresponding to the tearing of the SO_4 away from the Cu there.

Now this concentration in one compartment and weakening in the other compartment corresponds much more nearly than the first hypothesis does with what is observed, the truth being of course something intermediate between the two extremes. Anions and cations both travel, and so the total E.M.F. is shared in certain proportions between all the junctions: a proportion which can be calculated from a knowledge of the relative rates of migration: though the observations of the true migration-ratios may be complicated by molecular electric endosmose through the porous cell. Indeed, so far as I know at present, the migration velocity of SO_4 has not yet been very well directly determined.

But, although all this is valid on any view as to the nature of solution, the above mode of statement belongs to the old-fashioned view of chemical combination, the notion of metallic solution which was in vogue a few years ago. It may be well to state the matter also in terms of dissociation ideas.

The fashionable method is to think of the solvent as little more than a menstrum for a stream of metallic ions to saunter through. We may think of the zinc ions travelling into the liquid and some copper ions leaving it, the adjustment of the proportion between the free zinc and the freed SO_4 being a matter of subsequent diffusion. The solid zinc has, as it were, a certain vapour-tension or osmotic pressure P, the zinc of the zinc sulphate has a certain less solution-pressure p, less because it is there less concentrated than in the solid. Let there be N molecules per c.c. of solid and only n active molecules per c.c. of liquid, then the steady potential-difference between the solid and the liquid will be given by

$$\frac{dV}{dp} = \frac{1}{ne},$$

for in the present case $\frac{v_1 - u_1}{v_1 + u_1} = 1$, the anions being supposed stationary at that junction, since they cannot diffuse into the metal.

So putting $p = \rho RT = nmRT,$
we get

$$dV = \frac{m}{e} . RT \frac{dn}{n};$$

or

$$V - V' = R\epsilon T \log \frac{N}{n} = \cdot 114 \log_{10} \frac{N}{n} \text{ volt},$$

as the theoretical difference of potential between a metal and a solution of any one of its salts ; ϵ being written for the electrochemical equivalent m/e. Even if it is a foreign salt or an acid the potential-difference should be the same, provided there is equilibrium, *i. e.* no "local action."

Suppose now that one metal, with effective concentration N_1 (active ions per c.c.), dips into a solution of concentration n_1 ; and that this solution changes abruptly to another strength n_2 of the same solution, the circuit being completed by another metal N_2 dipping into that second solution ; then the whole E.M.F. of the cell, on the above hypothesis, is

$$E = R\epsilon T \left(\log \frac{N_1}{n_1} + \frac{u-v}{u+v} \log \frac{n_1}{n_2} + \log \frac{n_2}{N_2} \right)$$

$$= \cdot 114 \left(\log_{10} \frac{N_1}{N_2} + \frac{2v}{u+v} \log_{10} \frac{n_2}{n_1} \right) \text{ volt.}$$

One case of this is the one we have already considered at length, where only one metal is employed, that is, where $N_1 = N_2$, and the E.M.F. depends wholly on the solutions. The other chief case is the ordinary simple Volta cell, where $n_2 = n_1$, and where the theoretical E.M.F. depends solely on the two metals.

But now suppose that the two liquids are different, as well as the two metals ; there will now be altogether four ionic velocities and four real or virtual concentrations:—

$$u_1, v_1 ; \quad u_2, v_2 ; \quad n_1, n_2 ; \quad N_1, N_2.$$

So assuming that each solution acts like vacuum to the other, we have intrinsic velocities u_1 out and v_2 in across the liquid boundary of the first solution, or an E.M.F at that place of

$$\cdot 114 \frac{u_1 - v_2}{u_1 + v_2} \log n_1 \text{ volt};$$

together with

$$\cdot 114 \frac{v_1 - u_2}{v_1 + u_2} \log n_2 \text{ volt}$$

to be added to it, for the same boundary, considered from the point of view of the other solution.

Hence in this case, with two different metals,

$$\frac{\text{E}}{\cdot 114 \text{ volt}} = \log_{10} \frac{\text{N}_1}{\text{N}_2} - \frac{2v_2}{u_1 + v_2} \log_{10} n_1 + \frac{2v_1}{u_2 + v_1} \log_{10} n_2.$$

And the first term is usually the most important. For the special case where $\frac{v_2}{u_1} = \frac{v_1}{u_2} = \frac{1}{r}$, the two last terms reduce to the simple concentration expression already familiar, $\frac{2}{1+r} \log \frac{n_2}{n_1}$. The osmotic pressure ratio n_2/n_1 for the liquids may be a measure of their relative conductivities; but the solution-pressure ratio N_1/N_2 for the metals seems to be exponentially connected with their chemical activity, *e. g.* their heat of oxidation.

It is needless to point out that in the above equation most voltaic cells are included, at least if one is content to omit from consideration the inert non-migrational non-chemical junction of the two metals.

Some remarkable cells of this kind have been suggested, where the liquids overpower the metals, so to speak; none more so than a cell specified by Hittorf*, who says that by surrounding copper with a solution of a cyanide it can be made electropositive to zinc, so that copper shall go into solution and zinc shall be turned out, being deposited upon a zinc plate.

The scheme for this cell of Hittorf's is said to be

$$\text{Zn} \mid \text{ZnSO}_4 \mid \text{K}_2\text{SO}_4 \mid \text{KCN} \mid \text{Cu}.$$

The object of the K_2SO_4 is, as in many of these cells, to interpose a neutral harmless substance between two solutions which otherwise would precipitate each other.

* *Zeits. f. physical. Chem.* 1892, vol. x. p. 592.

Perhaps some one will set up a cell of this kind and verify the statement; though Hittorf's is very high authority.

CHAPTER VI.

Electron or Corpuscular View of Voltaic Action.

The theory which seeks to reduce the whole material universe to electric charges, their motions and interactions, constitutes a gigantic subject upon which it is scarcely rash to predict that a great deal will be said during the ensuing century.

Dr. Larmor (Phil. Trans.) seeks not only to explain physical phenomena in general by means of electrons, but also to explain the electron itself as an intrinsic æther strain. Prof. J. J. Thomson (Phil. Mag. *passim*, especially December 1899) adduces experimental facts in favour of the real and even isolated existence of such charged corpuscles, whether they be pure electrons or not.

Meanwhile the mode of expression in term of corpuscles is less ambitious than that in terms of electrons, though both are revolutionary and striking enough. Let us take it then as granted hypothetically that every material atom is built up of the same fundamental corpuscles, each with the usual ionic charge; a hydrogen atom being composed of about 500 of such corpuscles of opposite signs, a sodium atom of about 10,000, and a mercury atom of about 100,000 of them.

Let every electric current (except a displacement current in free space) be associated with a transfer of corpuscles, a convection of ionic charges, whether in liquids or in metals or in gases; the only difference being that in liquids the travelling corpuscle carries the whole atom with it, whereas in gases it can travel in some cases isolated from all the rest of an atom *, though in other cases it may be clogged with a considerable molecular aggregate of atoms †; while in solids presumably a corpuscle can only travel by being handed on from one atom

* J. J. Thomson, Phil. Mag. Dec. 1899.

† *Cf.* Chattock, Phil. Mag. Nov. 1899; also Rutherford, Townsend, Wilson, &c., Phil. Trans.

to the next: though a shift inside each atom will correspond to a polarisation current.

At an electrode the corpuscle leaves its ion to do the best it can, and passes on into the metal, each atom receiving it into its atomic grouping and instantaneously passing on an equal corpuscle to the next: or at least that is what I suppose to happen, consistently with the hypothesis. At a junction of two metals, then, a corpuscle which had formed part of an atom of one metal finds itself received and incorporated by a different kind of atom. This may clearly involve a gain or a loss of energy, and accordingly some propelling or opposing E.M.F. It may be expressed if we choose somewhat in the language of Helmholtz's hypothesis, that one metal attracts electricity more forcibly than another. If the excess energy takes the form of molecular agitation, such a procedure may entirely account for the Peltier and Seebeck phenomena.

I have spoken of it above (p. 383) as a futile kind of "transmutation": "transmutation" because a portion of the substance of say an iron atom enters into and becomes part of the substance of a copper atom; "futile" because there is no effective conversion of iron into copper, since the amount of each metal remains the same as before.

If corpuscles could be passed into a metal without being passed out again, there would be a kind of transmutation, and this is hypothetically done when a body is charged with negative electricity; it has gained corpuscles which do not belong to it. But the fact that they do not belong to it—being evidently there on sufferance and tending to escape as soon as possible—shows that there is no real transmutation: a charged sodium atom is not a mere sodium atom but a sodium monad ion; it has one negative corpuscle too many.

It might be thought that by giving to a metal charges of positive and negative electricity alternately some extra atoms could be built up: but unfortunately this is not consistent with the fact that, under known circumstances, the negatively charged corpuscle is the mobile ingredient; so that superposing a positive charge on a negatively charged atom is not adding something to the corpuscles already there, but is removing the extra negative corpuscle and another as well; taking the atom down therefore instead of building it up, and leaving a void which automatically fills itself up whenever it gets a chance. It is just possible that some kind of effective transmutation might be effected in an electric spark or arc with different materials on either side of it; like a substance passing over from one crystalline form to another.

Whether combination of atoms takes place across any junction when a current passes, appears to be merely a question of relative mobility of corpuscles and atoms. If the atoms are free to move, if it is easier to move an atom across the interface than to detach from it a corpuscle, then combination occurs, and the conduction is then necessarily on one side essentially electrolytic; and at a liquid-liquid junction there may be double atomic freedom and complete electrolysis. But if the corpuscles are more detachable than the atoms, then the conduction is metallic and no alloy is formed.

But now, pressing further the osmotic pressure view, if the metals can evaporate into a liquid they ought perhaps to be able to evaporate into each other, and so the E.M.F. at a junction ought to be calculable as

$$E = R\epsilon T\left(\quad\right) \log \frac{n}{n'},$$

where the $\frac{n}{n'}$ may be taken as the ratio of the metallic densities; but what shall we put into the brackets for the ionic velocities? Not ionic velocities at all, but corpuscular velocities, and these corpuscular velocities may differ by reason of the different viscosity or obstruction offered by each metal; which again perhaps may be proportional to the density; but the solution can hardly be treated as a weak one, or the gaseous laws as applicable.

If gaseous laws are applicable to the corpuscles inside metal, the value of their characteristic constant R will be 500 or 600 times the value for hydrogen; but inasmuch as their m, or ϵ their electrochemical equivalent, is 1/500 or 1/600 of the hydrogen value, the resulting $R\epsilon T$ comes out just as usual.

The viscosity or obstruction offered by metal is probably its electrical resistance; and so the corpuscular velocity in each may be proportional to the conductivity.

Conductivity in a liquid means greater mobility of the ions, conductivity in a metal may mean greater mobility of the corpuscles; hence perhaps the E.M.F. of a metallic junction might be something like this:

$$E = \cdot 114 \frac{k-k'}{k+k'} \log \frac{m/\rho}{m'/\rho'} \text{ volt,}$$

where the k are conductivities and the m atomic weights and the ρ densities.

I would suggest that the curious deterioration suffered by some thermo-electric piles with lapse of time, especially those used for generating a fair amount of power, has never so far as I know attracted much attention or been thoroughly investigated. It may be merely mechanical, due to alternate heating and cooling, but it may possibly be due to something of greater interest.

The corpuscles clinging to atoms are rather like ions clinging to solvent; so perhaps in a metal the negative corpuscles move with velocity

$$u = u_1 \left(\frac{1}{Ne} \frac{dp}{dx} + \frac{dV}{dx} \right)$$

where the intrinsic velocity u_1 is large; while the positive corpuscles move with velocity

$$v = v_1 \left(\frac{1}{Ne} \frac{dp}{dx} - \frac{dV}{dx} \right),$$

their intrinsic velocity v_1 being very small.

But in a homogeneous metal itself these speeds u and v are perhaps equal; indeed where there is no potential gradient or driven current they must be equal.

Are they equal at a metallic junction too?, or is anything kinetic occurring there at all?

There must be a stress or polarisation or double layer or something statical at such a junction; and a statical condition can be simulated by kinetic equilibrium. What are the arguments?

When a zinc surface is illuminated with ultra-violet light J. J. Thomson found that it does not throw off corpuscles unless it is negatively electrified. The ultra-violet light may be said to detach them, but it takes electrification to throw them off.

So at a junction of two metals when no current is being passed it may be that there is no interchange of substance; but the interchange may or must begin as soon as a current flows; and if the A aggregation has more potential energy then the B aggregation there will be a force urging the corpuscles to flow from A to B.

In a gas the negative corpuscles are by far the most mobile; indeed the positive corpuscles seem unable to move except with their atoms electrolytically. It may be that even inside a metal the negative corpuscles are likewise by far the most mobile, being readily handed on from one atom to the next, this process constituting metallic conduction; while the positive corpuscles remain attached to their atoms, which except in an electrolyte or a substance with some electrolytic properties are not subject to locomotion at all. Assuming this mobility of corpuscles of one sign solely, they are urged along a gradient of temperature in a metal for two reasons: one a mechanical force equivalent to a dp/dx acting on an element of volume, or $1/n$ of this per corpuscle; the other an electrical force dV/dx acting on its electric charge e.

When there is no current these forces must balance. So

$$\frac{1}{n}\frac{dp}{dx} - e\frac{dV}{dx} = 0,$$

or

$$dp = ned\mathrm{V}.$$

Now proceeding as usual without compunction to utilise the gas-analogy

$$p = nm\mathrm{RT},$$

we have

$$ned\mathrm{V} = m\mathrm{R}d(n\mathrm{T})$$

or

$$d\mathrm{V} = m/e \,.\, \mathrm{RT}d\log(n\mathrm{T}).$$

This is the E.M.F. that acts along the gradient of temperature, giving rise to a reversible evolution of heat and representing the coefficient of the Thomson effect, σ, called the specific heat of electricity in a metal, such that the work done in transferring a charge q up a difference of temperature dT is $qdV = q\sigma dT$, or

$$\sigma = \frac{m}{e} RT \frac{d}{dT} \log (nT).$$

Hence for a circuit of two metals a and b

$$\sigma_a - \sigma_b = R\epsilon T \frac{d}{dT} \log \frac{n_a}{n_b}.$$

And by ordinary thermodynamics this is related to the Peltier effect at the junction and to the resultant E.M.F. in the circuit by the equations

$$\int \frac{\sigma_a - \sigma_b}{T} dT = \frac{dE}{dT} = \frac{\Pi}{T};$$

therefore

$$\Pi = R\epsilon T \log \frac{n_a}{n_b}.$$

Now n represents the number of active corpuscles in unit volume, and this number is a function of the temperature; and its rate of change probably depends on the number already there and on a coefficient characteristic of the metal; $\frac{dn}{n} = \alpha\, dT$, say; therefore let $n_a = ae^{\alpha T}$, $n_b = be^{\beta T}$.

In substances whose conductivity increases with temperature α will be positive, but in metals it is usually negative, and at absolute zero the conductivity would rise to a maximum value corresponding to the maximum corpuscular concentration a or b.

So

$$\Pi = R\epsilon T \log \frac{a}{b} + (\alpha - \beta) R\epsilon T^2;$$

whence the neutral point is

$$T_0 = \frac{\log b - \log a}{\alpha - \beta},$$

and

$$\sigma_a = \alpha R\epsilon T = k_a T,$$

and

$$\Theta_a = \tfrac{1}{2}\alpha R\epsilon (T_1^2 - T_2^2),$$

and

$$E = (\alpha - \beta) R\epsilon (T_1 - T_2)\left(T_0 - \tfrac{1}{2}(T_1 + T_2)\right),$$

or

$$P_a = \alpha R\epsilon (T_0 - \overline{T}).$$

So that our hypothesis does fairly well to express thermoelectric facts; but, unless the opposite sign of corpuscles is attended to, it does not account for the fact that the coefficient σ is positive in some metals and negative in others; though it suggests an opposite sign for σ in metals and electrolytes, because α changes sign along with the temperature-coefficient of conductivity.

To divide metals into two classes we must not suppose that the negative corpuscles do all the travelling: if they did, moreover, the Hall effect, the magnetic influence on the current, corresponding to the magnetic curvature of cathode rays, would be much greater than it is. The facts that the Hall effect is small, and that it has different signs in different metals, suggest, as J. Larmor points out, that the speed of the negative and positive corpuscles or electrons is nearly equal, but that in some metals the positive travel slightly faster, in other metals the negative: the difference being subordinate and secondary.

It will turn out on this view that the "specific heat of electricity" σ, the coefficient of the Thomson effect, has the same sign as the coefficient of magnetic curvature of current, the Hall effect. For proceed with our theory of metallic conduction treated as a convection of charged corpuscles of both kinds, with u/v the Hittorfian migration ratio of negative to positive corpuscular velocity, under the action of unit force; and write down as in electrolysis that the resultant motion must be equal in opposite directions,

$$u\left(\frac{1}{n}\frac{dp}{dx} + e\frac{dV}{dx}\right) = v\left(\frac{1}{n}\frac{dp}{dx} - e\frac{dV}{dx}\right),$$

or

$$\frac{dV}{dp} = \frac{1}{ne}\frac{u-v}{u+v},$$

or

$$dp = ne\frac{u+v}{u-v}dV.$$

Then the various thermoelectric quantities get written exactly as above, except that the new factor $\frac{u-v}{u+v}$ must be everywhere introduced.

It is worth noticing that

$$ne(u+v) = \frac{C/A}{dV/dx} = \frac{\text{current-density}}{\text{potential-gradient}} = \text{specific conductivity} = \frac{1}{\rho}$$

wherefore

$$\sigma = \frac{dV}{dT} = \rho(u-v)\frac{dp}{dT}$$

$$= \rho m R(u-v)\frac{d}{dT}(nT).$$

This form is instructive as showing that, other things being equal, highly resisting substances are likely to form the best thermoelectric materials. The metals of high conductivity have, in fact, feeble thermoelectric power; metals like bismuth and antimony are poor conductors; and the contact force between insulators, or still better between an insulator and a conductor, may be enormous.

But for most purposes the previous form of expression is handier, so introducing the migration ratio $r = u/v$ we get

$$\sigma = R\epsilon T\frac{1-r}{1+r}\frac{d\log(nT)}{dT};$$

and so on for Π and E as before when two metals are employed, only with the new relative migration factor which permits ready change of sign; and it must be remembered that the corpuscular migration ratio r, as well as the corpuscular detachment number n, may differ in the two metals.

Chapter VII.

Helmholtz's View of Electric Mass-Affinity.

The doctrine that matter has a specific attraction for electricity * leads to many of the same conclusions as the supposition that positive or negative electricity inside a substance has a real specific heat, and the two modes of statement may be considered together; but we may speak in the language of Helmholtz's hypothesis, as being the more important of the two, indeed the only one seriously upheld as a physical reality. As has been suggested above, p. 389, it is to some extent ambiguous, however; for it is not certain whether the affinity or attraction is effective at a metal-æther free surface or at a metal-metal junction. As usually stated, the attraction is supposed to be an affair of mass or bulk and not of surface, though no doubt any such internal volume electric stress would give rise to something analogous to electrical surface-tension †; but the whole may be really and essentially superficial, having to do with the free surface only; and in that case it is difficult to distinguish it from chemical affinity, unless it can truly be shown to be independent of the presence and nature of surrounding adjacent matter. Even so it may be considered as an early and crude statement of certain consequences of the electron theory of matter, rather than as a distinct and independent theory (*cf.* J. Larmor, Phil. Trans. 1897, vol. cxc. p. 271).

However it be regarded, it is plain that the effect of an attraction between matter and electricity must display itself at a transition or boundary layer, and although it has been customary to concentrate attention on the metallic (*e.g.* the copper-zinc) junction in this connexion, it is by no means probable that that is so efficient or complete a transition as is the free air or æther surface of either metal; nevertheless these surface actions have secured too little attention.

The proofs that have been given, for instance, of the value

* For references see Phil. Mag. October 1885, p. 377.

† Just as Laplace's K gives rise to capillarity.

of the Peltier effect, and its connexion with contact force, or boundary force as we might now call it, are accustomed more or less completely to ignore the air boundaries; employing an open circuit with the object of avoiding obvious neutralising or compensating forces at other junctions, but forgetting that opening a circuit necessarily introduces one *more* boundary, not one *less*.

Dr. Larmor has been good enough to tell me of a "proof" in Parker's 'Thermodynamics' (1894 edition) that on Helmholtz's view the Peltier coefficient at a metallic junction is proportional to the rate of change of the junction force t[illegible] with temperature; a proof somewhat on the same lines as that given independently by Lord Kelvin in the Proc. Roy. Soc. Edin. vol. xxii. page 118, Session 1897–8; but neither proof in reality establishes anything specially limited to the metallic junction; the whole of the junctions are really laid under contribution, a fact which is half recognised by Lord Kelvin, though not by Mr. Parker.

In order to save other people from working through these proofs I may reword them in a simpler and briefer fashion, using language appropriate to the mass-affinity doctrine, but incidentally throwing in the symbol σ so as to include also the essence of the specific heat idea.

Let a charge of electricity given to a metal be held there by an intrinsic attraction or mass-affinity; it will have a special affinity-energy in addition to the energy of its ordinary electrical interactions with other charges: call this surplus energy, at the absolute temperature θ,

$$u=u_0+\int\sigma d\theta.$$

The corresponding entropy is

$$s=s_0+\int\frac{\sigma d\theta}{\theta}=s_0+\int\frac{du}{\theta},$$

so that

$$du=\sigma d\theta=\theta ds.$$

Consider an open circuit of two active metals and one neutral substance, *e. g.* Zn, Cu, and Pt, or 1, 2, and 0, with air-gaps forming condensers as in the diagram ; the condensers being charged to the potentials V_1 and V_2 respectively by the differential electric affinities at work.

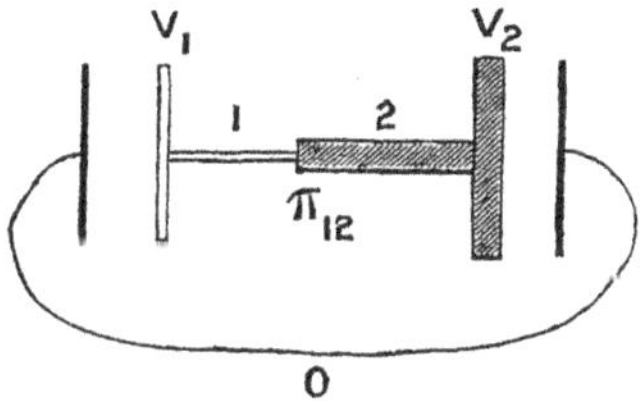

Now transfer unit charge reversibly from one condenser to the other, *i. e.* from potential V_1 to V_2, by shifting the neutral plates ; the work done, equivalent to loss of energy, has usually been written as

$$V_2 - V_1 = \Pi_{12} + u_1 - u_2,$$

but it ought to be written

$$V_2 - V_1 = \Pi_{01} + u_1 + \Pi_{12} - u_2 + \Pi_{20}.$$

The total entropy change is zero,

$$\frac{\Sigma\Pi}{\theta} + s_1 - s_2 = 0,$$

or, writing Π for $\Sigma(\Pi)$,

$$\Pi = \theta(s_2 - s_1),$$

wherefore

$$V_2 - V_1 = u_1 - u_2 - \theta(s_1 - s_2),$$

or

$$V + u - \theta s = \text{constant}.$$

But

$$u - u_0 = \int \theta ds = \theta s - \int s d\theta,$$

so

$$V - \int s d\theta = \text{constant},$$

or

$$\frac{dV}{d\theta} = s,$$

or

$$\frac{\Pi}{\theta} = s_2 - s_1 = \frac{d}{d\theta}(V_2 - V_1);$$

the result to be proved.

But there is nothing here more than has been admitted in this paper over and over again.

If surface boundary forces are supposed zero, if electricity can enter or leave the free surface of a metal without aid or obstruction, then $V_1 - V_2$ must represent the only remaining force, viz. that at the metallic junction; and the only Π will be situated there; but if there are other boundary forces, then $V_1 - V_2$ represents the whole E.M.F. of the circuit, and not the localised junction force at all, and Π_{12} is not necessarily the sole Peltier effect. The metallic junction force may still be measured by Π_{12}; the combined entropy of the boundary forces is $\Sigma\Pi/\theta$; and this entropy is proved equal to $dE/d\theta$, a remarkable and interesting but long known fact.

Returning to the theory of Helmholtz, which was stated first so long ago as 1847, and elaborated to some extent in his brilliant "Faraday lecture" of 1881, it is undoubtedly very important; it may be regarded as the precursor of, and therefore to some extent superseded by, the electron or corpuscular theory of matter: a theory which represents the natural outcome, in the light of recent experiments, of that "atomic" view of electricity which, started by Clerk Maxwell in an inspired phrase, adopted and expanded by Dr. Johnstone Stoney and others, is now in the preceding pages attempted to be applied, in conjunction with ideas derived from the treatment of osmotic pressure by physical chemists, to an explanation of thermoelectric action on the gaseous analogy.

Made in the USA
Monee, IL
04 November 2026

67387479R10036